Hen Eggs

Their Basic and Applied Science

Edited by

Takehiko Yamamoto, Ph.D.
Professor Emeritus
Osaka City University
Osaka, Japan

Lekh Raj Juneja, Ph.D.
Director (R&D)
Central Research Laboratories
Taiyo Kagaku Co., Ltd.
Yokkaichi, Mie, Japan

Hajime Hatta, Ph.D.
Assistant General Manager (R&D)
Central Research Laboratories
Taiyo Kagaku Co., Ltd.
Yokkaichi, Mie, Japan

Mujo Kim, Ph.D.
Managing Director (R&D)
Central Research Laboratories
Taiyo Kagaku Co., Ltd.
Yokkaichi, Mie, Japan

CRC Press
Boca Raton   New York   London   Tokyo

Acquiring Editor:	Marsha Baker
Editorial Assistant:	Jean Jarboe
Project Editor:	Debbie Didier
Marketing Manager:	Susie Carlisle
Direct Marketing Manager:	Becky McEldowney
Cover design:	Dawn Boyd

Library of Congress Cataloging-in-Publication Data

Hen eggs : their basic and applied science / edited by Takehiko
    Yamamoto . . . [et al.].
        p.   cm.
    Includes bibliographical references and index.
    ISBN 0-8493-4005-5
    1. Eggs.   2. Eggs--Composition.   I. Yamamoto, Takehiko.
SF490.H46   1996
598.6′17--dc20   96-38268
                                                                                        CIP

This book contains information obtained from authentic and highly regarded sources. Reprinted material is quoted with permission, and sources are indicated. A wide variety of references are listed. Reasonable efforts have been made to publish reliable data and information, but the author and the publisher cannot assume responsibility for the validity of all materials or for the consequences of their use.

Neither this book nor any part may be reproduced or transmitted in any form or by any means, electronic or mechanical, including photocopying, microfilming, and recording, or by any information storage or retrieval system, without prior permission in writing from the publisher.

All rights reserved. Authorization to photocopy items for internal or personal use, or the personal or internal use of specific clients, may be granted by CRC Press, Inc., provided that $.50 per page photocopied is paid directly to Copyright Clearance Center, 27 Congress Street, Salem, MA 01970 USA. The fee code for users of the Transactional Reporting Service is ISBN 0-8493-4005-5/97/$0.00+$.50. The fee is subject to change without notice. For organizations that have been granted a photocopy license by the CCC, a separate system of payment has been arranged.

The consent of CRC Press does not extend to copying for general distribution, for promotion, for creating new works, or for resale. Specific permission must be obtained in writing from CRC Press for such copying.

Direct all inquiries to CRC Press, Inc., 2000 Corporate Blvd., N.W., Boca Raton, Florida 33431.

© 1997 by CRC Press, Inc.

No claim to original U.S. Government works
International Standard Book Number 0-8493-4005-5
Library of Congress Card Number 96-38268
Printed in the United States of America 1 2 3 4 5 6 7 8 9 0
Printed on acid-free paper

EDITORIAL BOARD

Chairman	Takehiko Yamamoto, Ph. D.	Osaka City University
Editors	Lekh Raj Juneja, Ph. D.	Central Research Laboratories Taiyo Kagaku Co., Ltd.
	Hajime Hatta, Ph. D.	Central Research Laboratories Taiyo Kagaku Co., Ltd.
	Mujo Kim, Ph. D.	Central Research Laboratories Taiyo Kagaku Co., Ltd.
Board Office	Central Research Laboratories Taiyo Kagaku Co., Ltd. 1-3 Takaramachi, Yokkaichi Mie 510 Japan	
	Tel: +81-593-47-5400 Fax: +81-593-47-5417	

CONTRIBUTORS

Akira Seko, Ph.D [a]
Researcher
Central Research Laboratories
Taiyo Kagaku Co., Ltd.

Hajime Hatta, Ph.D
Assistant General Manager
Central Research Laboratories
Taiyo Kagaku Co., Ltd.

Hidehisa Takahashi, Ph.D [b]
Manager
Central Research Laboratories
Taiyo Kagaku Co., Ltd., Japan

Hidetoshi Sugino
Group Leader
Central Research Laboratories
Taiyo Kagaku Co., Ltd.

Hisham R. Ibrahim, Ph.D [c]
Group Leader
Central Research Laboratories
Taiyo Kagaku Co., Ltd.

Kazuhiko Hirano
Assistant Manager
Central Research Laboratories
Taiyo Kagaku Co., Ltd.

Ken Tsuda
Group Leader
Central Research Laboratories
Taiyo Kagaku Co., Ltd.

Lekh R. Juneja, Ph.D
Director
Central Research Laboratories
Taiyo Kagaku Co., Ltd.

Makoto Ozeki
Group Leader
Central Research Laboratories
Taiyo Kagaku Co., Ltd.

Mamoru Koketsu, Ph.D
Group Leader
Central Research Laboratories
Taiyo Kagaku Co., Ltd.

Miguel A. Gutierrez, Ph.D
Researcher
Central Research Laboratories
Taiyo Kagaku Co., Ltd.

Mikio Kobayashi
Researcher
Central Research Laboratories
Taiyo Kagaku Co., Ltd.

Shigemitsu Akachi, Ph.D
Assistant Manager
Central Research Laboratories
Taiyo Kagaku Co., Ltd.

Takashi Hagi
Assistant Manager
Central Research Laboratories
Taiyo Kagaku Co., Ltd.

Takehiko Yamamoto, Ph.D
Professor Emeritus
Osaka City University

Teruhiko Nitoda [d]
Researcher
Central Research Laboratories
Taiyo Kagaku Co., Ltd.

Tsutomu Okubo
Assistant Manager
Central Research Laboratories
Taiyo Kagaku Co., Ltd.

PRESENT ADDRESS

[a] Sasaki Research Institute, Tokyo
[b] Coki Co., Ltd., Tsu
[c] Kagoshima University, Kagoshima
[d] Okayama University, Okayama

PREFACE

The egg of a chicken is an encapsulated next generation which, under suitable conditions, turns into a chicken. This event is achieved utilizing only the inner components of the shell with the exception of changing air. In other words, an egg contains all kinds of substances necessary for a chicken to hatch. Unlike mammals, birds lack a lactation system, but are endowed with the ability to pick their feed in a day after hatching. Also, the hatched chicken has antibodies to protect it from various infectious diseases. It could be said that, for an egg to turn into a chicken, a precision creature, in a closed system, the event is none other than a mystery, even though this event would have been performed according to a program genetically encoded on the gene of the embryo.

Bird eggs as well as cow's milk have been used as food by human beings since prehistoric time. Nowadays, the daily consumption of hen eggs in the world is estimated at nearly a half billion, and most of them are for domestic use. However, recently, considerable amounts of hen eggs have been processed to be a material for the food industry or as confectionery. Hen egg protein has been recognized as one of the highest quality proteins. Particularly, the egg yolk protein has been evaluated to be superior to cowmilk protein in digestibility as well as amino acid composition. Hen eggs have also been evaluated as a source of several vitamins and minerals, and, thus, daily intake of hen eggs is known to supply several vitamins and minerals in nearly the amount recommended as our daily allowance. As for the chemistry of the protein of hen eggs and their biological, nutritional, biophysicochemical properties and bioeconomic values, several professional books have been published (e.g., Warner, R. C., *The Proteins*, Neurath, H. and Bailey, K., Eds., Academic Press, Inc., 1959, Vol. 2, p. 436~485, Romanoff, A. L. and Romanoff, A. J., *The Avian Egg*, John Wiley & Sons, Inc., New York, 1949, and Burley, R. W. and Vadehra, D. V., *The Avian Egg- Chemistry and Biology*, John Wiley & Sons, Inc., New York, 1989).

Our research group has made a great effort to find new applications for egg white, egg yolk, and whole egg based on their characteristic properties. The research of seeking biologically or biochemically active substances in the hen eggs has also been performed and it was found that the hen egg is an excellent source of carbohydrates, such as, sialic acid, sialyloligosaccharides, and polyunsaturated fatty acids, such as, arachidonate, and docosahexaenoate, attached to phospholipids. On the other hand, the immune system of the hen egg was investigated and it was revealed that the antibodies induced in the hen are transferred to the egg yolk in a concentrated state in the egg laid by the hen. It was also found that the antibodies, whose induction is difficult in mammals, are readily obtainable through the immune system of the hen and egg. Some of the research results have, thus, been applied practically.

This year is the fiftieth anniversary of the foundation of Taiyo Kagaku Co., Ltd., to which our Central Research Laboratories belong. To dedicate this anniversary, our research results published in various academic journals or authorized as a patent are compiled by several editors and published as <u>Hen Eggs-Their Basic and Applied Science</u> by the courtesy of CRC

Press. We hope this book will be useful as a reference not only for researchers and engineers working in the poultry industry, but also for those in biology and biochemistry. On this occasion, we would like to express our sincere thanks to Mr. Nagataka Yamazaki, President of Taiyo Kagaku Co., Ltd., for his very positive support for publishing this book.

Takehiko Yamamoto, Ph. D.
Chairman Editorial Board

TABLE OF CONTENTS

CHAPTER 1   1
STRUCTURE OF HEN EGGS AND PHYSIOLOGY OF EGG LAYING
T. OKUBO, S. AKACHI, AND H. HATTA

CHAPTER 2   13
GENERAL CHEMICAL COMPOSITION OF HEN EGGS
H. SUGINO, T. NITODA, AND L.R. JUNEJA

CHAPTER 3   25
NUTRITIVE EVALUATION OF HEN EGGS
M. A. GUTIERREZ, H. TAKAHASHI, AND L.R. JUNEJA

CHAPTER 4   37
INSIGHTS INTO THE STRUCTURE-FUNCTION RELATIONSHIPS OF
OVALBUMIN, OVOTRANSFERRIN, AND LYSOZYME
HISHAM R. IBRAHIM

CHAPTER 5   57
EGG YOLK PROTEINS
L.R. JUNEJA AND M. KIM

CHAPTER 6   73
EGG YOLK LIPIDS
L.R. JUNEJA

CHAPTER 7   99
GLYCOCHEMISTRY OF HEN EGGS
M. KOKETSU

CHAPTER 8   117
CHEMICAL AND PHYSICOCHEMICAL PROPERTIES OF HEN EGGS AND
THEIR APPLICATION IN FOODS
H. HATTA, T. HAGI, AND K. HIRANO

CHAPTER 9   135
ENZYMES IN UNFERTILIZED HEN EGGS
A. SEKO, L.R. JUNEJA, AND T. YAMAMOTO

CHAPTER 10   145
CELL PROLIFERATION-PROMOTING ACTIVITIES IN UNFERTILIZED EGGS
A. SEKO AND L.R. JUNEJA

CHAPTER 11   151
EGG YOLK ANTIBODY IGY AND ITS
APPLICATION
H. HATTA, M. OZEKI, AND K. TSUDA

CHAPTER 12   179
MICROBIOLOGY OF EGGS
M. KOBAYASHI, M. A. GUTIERREZ, AND H. HATTA

INDEX   193

Hen Eggs

Their Basic and Applied Science

Chapter 1

STRUCTURE OF HEN EGGS AND PHYSIOLOGY OF EGG LAYING

T. Okubo, S. Akachi, and H. Hatta

TABLE OF CONTENTS

I. Introduction
II. Structure of Hen Eggs
    A. Eggshell
        1. Cuticle
        2. Shell Stratum
        3. Shell Membrane
    B. Egg Albumen
        1. Thick and Thin Albumens
        2. Chalaziferous Layer and Chalazae Code
    C. Egg Yolk
        1. Vitelline Membrane
        2. Structure of Yellow Yolk
III. Physiology of Egg Laying
    A. Ovary
        1. Development of the Ovary
        2. Rapid Growth of Follicles in the Ovary
    B. Process of Egg Laying
        1. Infundibulum
        2. Albumen-secreting Portion
        3. Isthmus
        4. Uterus
        5. Vagina
References

I. INTRODUCTION

A laid avian egg is a potential life precursor to the next generation of birds. Humans have utilized hen eggs as a nutritional food since ancient times. The chemical changes occurring inside an egg during its hatching are mysterious and dramatic. However, it is recognized that there is no significant difference in chemical composition and nutritional value between fertilized and unfertilized eggs, as far as they are fresh. An egg is considered to be a chemical and nutritional store for a potential new life. This chapter describes, briefly, the structure of hen eggs and the process of laying eggs to aid in understanding of subjects in the subsequent chapters.

II. STRUCTURE OF HEN EGGS

A hen egg is composed of three main parts: shell, albumen (egg white), and yolk. The yolk is surrounded by an albumen layer, and this structure is covered by a hard eggshell

(Figure 1). The weight of a hen egg and the weight distribution of the three parts differ considerably depending on the kind of hens and their age. The eggs of a white leghorn weigh from 50 to 63 g and the weight distributions of shell, albumen, and yolk are in the range of 9-11%, 60-63%, and 28-29%, respectively.

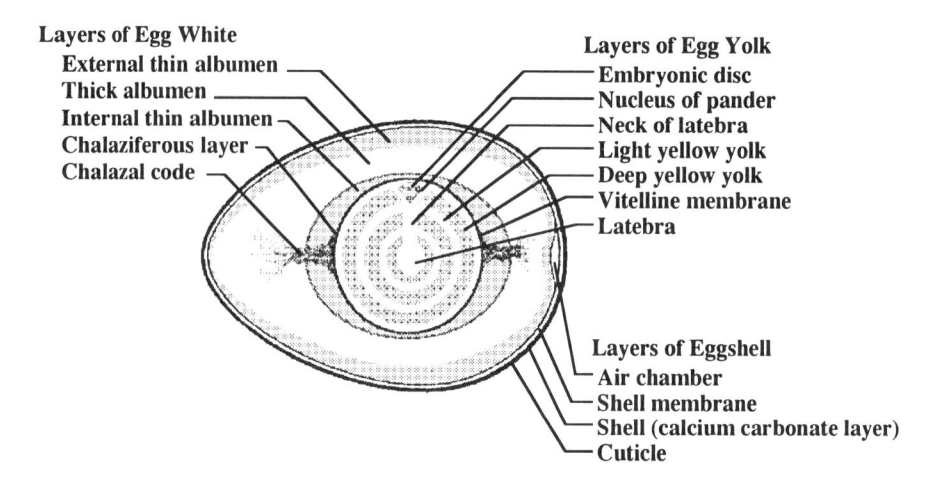

Figure 1. Structure of the egg.
Modified from Romanoff and Romanoff (1949) [16] by permission of John Wiley & Sons, Inc., New York.

A. EGGSHELL

An eggshell is composed of a thin film of cuticle, a calcium carbonate layer, and two shell membranes. Figure 2 is an electron micrograph of a tangential section of the eggshell. The microstructure of the eggshell is depicted in Figure 3. There are funnel-shaped small holes called pore canals on the surface of the shell for gas exchange. The pore canals are scatteringly located between the palisade layers of the shell, directed to the exterior. The diameter of the pore canal ranges from 10 to 30 μm. An egg has about 10,000 pore canals on the shell surface. The pore canal allows air and moisture to pass through, but does not allow liquid water.

1. Cuticle

The cuticle, the most external layer of eggs, is about 10 μm thick and covers the pore canals. It protects the egg from moisture and invasion of microorganisms to a certain extent [1]. The cuticle, of course, permits the exchange of gas in the egg. The cuticle is removed from the shell easily by soaking eggs in either weak acid solutions or metal chelator-containing solutions or by washing with water [2]. Therefore, washing of eggs often facilitates bacterial invasion of the egg.

The cuticle consists of protein, small amounts of carbohydrates, and lipids [3-5]. As shown in Table 1, it is interesting that the cystine content of the cuticle is almost half of the cystine content of shell membrane. The carbohydrates of cuticle are galactose, mannose, fructose and hexosamine, and the molar ratio of hexosamine to neutral sugar has been reported to be 1.0 to 0.93. In the case of lipids, Suyama and his colleagues reported that the ratio of neutral lipids to polar lipids was 6:1, while the ratio to that of egg yolk was 2:1 [6].

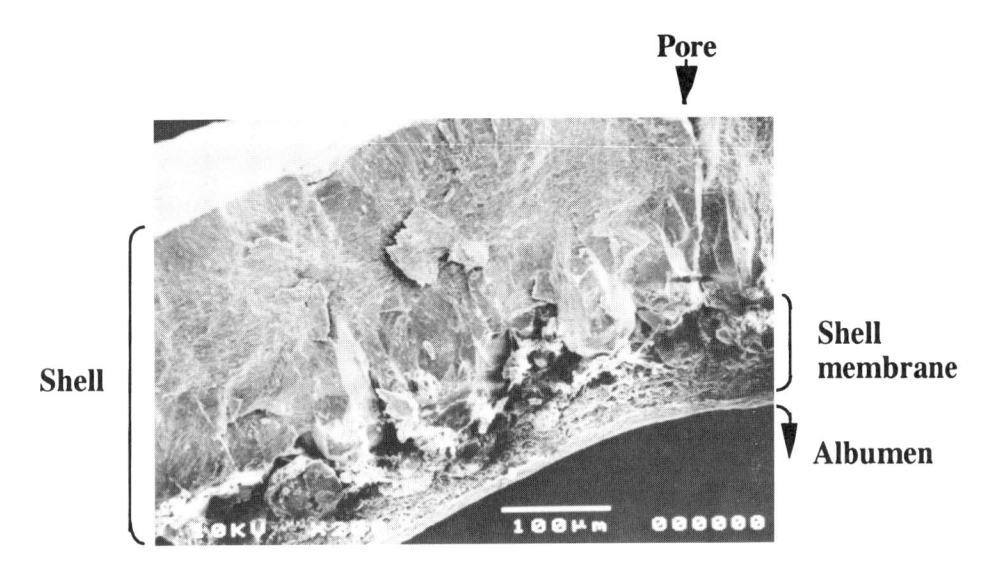

Figure 2. Scanning electron microscopic photograph of the shell of a hen egg.

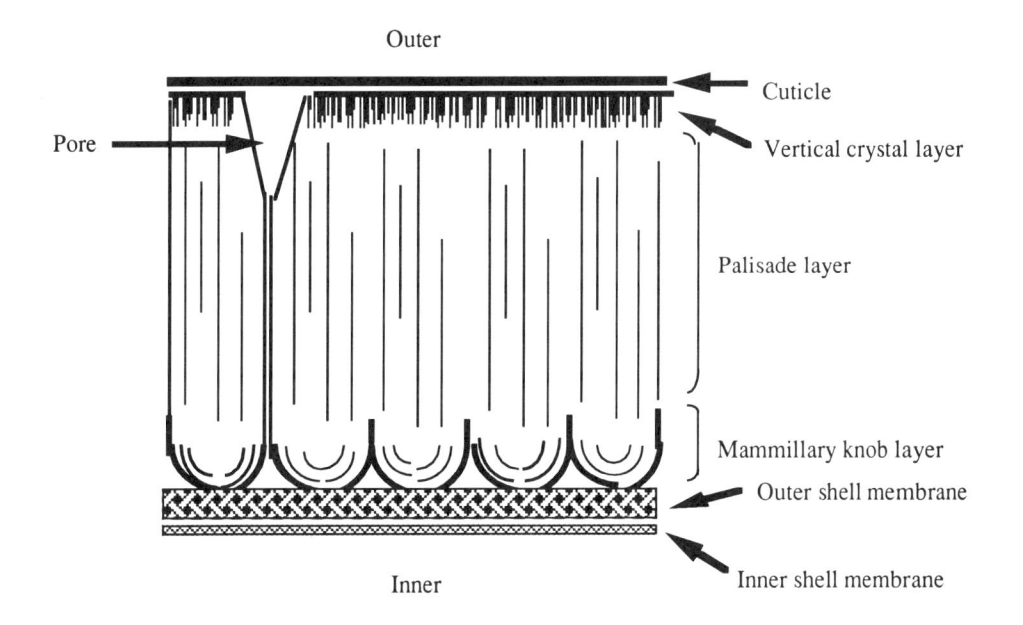

Figure 3. An illustrative representation of the hen eggshell.

Table 1
Amino Acid Composition
of Cuticle and Shell Membrane

Amino acid	Cuticle	Shell membrane
Aspartic acid	8.51	8.67
Threonine	5.75	6.04
Serine	6.02	5.97
Glutamic acid	11.92	13.56
Glycine	9.65	6.19
Alanine	3.82	3.47
Cystine	4.01	8.89
Valine	3.49	6.75
Methionine	3.51	3.28
Isoleucine	4.11	3.12
Leucine	4.30	4.72
Tyrosine	4.81	2.31
Phenylalanine	2.22	2.54
Lysine	5.07	4.37
Histidine	0.92	2.56
Arginine	6.01	6.87
Proline	4.85	8.91

Amino acids, g/100 g sample, dried

2. Shell stratum

The egg shell consists of the vertical crystal layer, palisade layer, and mammillary knob layer with an average thickness of 5 μm, 200 μm, and 110 μm, respectively (Figure 3). The components are 95% inorganic substance, 3.3% protein, and 1.6% moisture. Calcium carbonate is the major component of the inorganic substances. The vertical crystal layer consists of short, thin crystals running in a vertical direction of the shell. This layer involves small vesicles, but its structure remains unsolved [7]. The palisade layer is very dense and hard. Its crystalline structure is formed by calcification of calcium carbonate containing a small amount of magnesium, which constructs a spongy matrix together with collagen. The palisade layer is, thus, called the spongy layer. Each mammillary knob has a core and is in contact with the outer shell membrane. The dense distribution of the mammillary knobs on the shell membrane serves to harden the shell.

3. Shell membrane

The eggshell membrane is composed of inner and outer membranes. Their structure is like entangled threads or randomly knitted nets (Figure 4). This structure is important in obstructing invading microorganisms by catching them in the meshwork. The thickness of outer and inner membrane is about 50 µm and 15 µm, respectively. They consist of 70% organic substance, 10% inorganic substance, and 20% moisture. The main organic constituent is protein with a small amount of lipids and carbohydrates. The lamellar structure of the shell membrane is constructed of a thin insoluble fibrous layer of protein with numerous meshworks. Its amino acid composition is shown in Table 1. The cystine content of the shell membrane is twice as much as that of the cuticle. Candish and his colleagues confirmed the presence of L-5-hydroxylysine (1-2 moles/105 g protein) in the acid hydrolysates of shell membrane, suggesting that the shell membrane contains collagen [8]. Wong and his colleagues also found the presence of collagen-like proteins (collagen types I and V) in the inner and outer shell membranes [9]. Furthermore, Arias and his colleagues observed in an experiment using immunochemical analysis that the egg shell membrane contains Type X collagen [10]. Desmosine, isodesmosine, and nonelastin protein have also been found as the cross-linking amino acids in the eggshell membrane [11-13]. Taking the above findings into consideration, it is highly likely that keratin, collagen or elastin like protein exist in the shell membrane. Suyama and his colleagues reported that the lipid content of the shell membrane was 1.35% and the ratio of neutral lipids to complex lipids was 86:14 [6]. The main components of the complex lipids are sphingomyelin (63%) and phosphatidylcholine (12%).

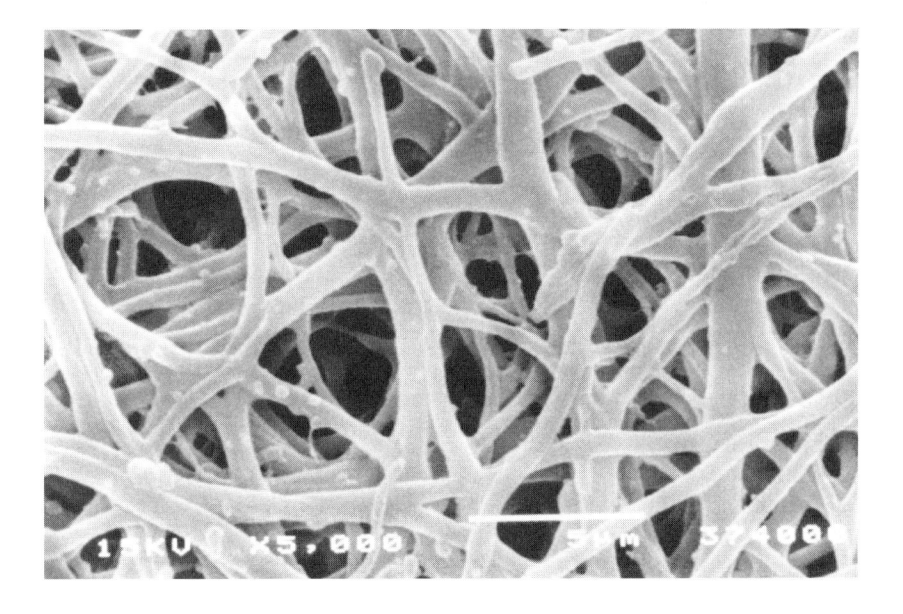

Figure 4. Scanning electron microscopic photograph of shell membrane.

B. EGG ALBUMEN

The egg albumen portion consists of thin and thick albumen and a chalaziferous layer. Thick albumen is sandwiched between outer and inner thin albumen.

1. Thick and thin albumens

The viscosity of thick albumen is much higher than that of thin albumen. This high viscosity is due to the high content of ovomucin, its concentration being four times as much as that of thin albumen. In fresh eggs, thick albumen covers the inner thin albumen and the chalaziferous layer, keeping the egg yolk in the center of the egg. Thick albumen is also in direct contact with the shell membrane.

2. Chalaziferous layer and chalazae code

The chalaziferous layer is a fibrous layer and directly covers the entire egg yolk. In the long axis of the egg, the chalaziferous layer is twisted at both sides of the yolk membrane, forming a thick rope-like structure named the chalazae code. This code is twisted clockwise at the sharpened end of the egg and counterclockwise at the opposite end. The chalazae code stretches into the thick albumen layer to both sides, thus, the egg yolk is suspended in the center of egg.

C. EGG YOLK

The egg yolk is encircled by a vitelline membrane. Structurally, the inner content consists of yellow yolk and white yolk. The white yolk originates from the white follicle which matures in the ovary. It has been reported that the weight of the white yolk is less than 2% of the total egg yolk. The yellow yolk is composed of layers of alternate light and deep yellow yolks. Morphologically, several structures are seen in egg yolk (latebra, neck of latebra, nucleus of pander, and embryonic disc), which originate from white yolk. Latebra is located in the center of the egg yolk and is connected with the nucleus of pander through a tube-like thread, named "neck of latebra." The embryonic disc (2-3 mm in diameter) in the nucleus of pander is the place for the embryo to develop.

1. Vitelline membrane

The vitelline membrane is composed of an inner layer, continuous membrane, and outer layer. Their thickness are 1.0-3.5, 0.05-0.1, and 3-8.5 µm, respectively. The inner and outer layers are three dimensionally mesh-worked structures consisting of fibers with diameters of 200-600 and 15 nm, respectively. On the other hand, the continuous membrane is a piled, sheet-like structure consisting of granules with an estimated diameter of 7 nm. The weight of the vitelline membrane is 51 mg per egg in average, and its solid matter is 20-30%. The solid matter of the vitelline membrane consists of protein (87%), lipids (3%), and carbohydrates (10%). It also contains DNA and RNA [14, 15]. Because of its amino acid composition, the protein of the vitelline membrane is not classified in the categories of collagen, keratin, or elastin.

2. Structure of yellow yolk

Obtaining the material through the mother hen's blood stream, the white follicle rapidly accumulates yellow yolk to become a yellow follicle in about 9 days before it is ovulated into the oviduct. Yolk consists of two types of lipoprotein emulsion, the deep yellow yolk and the light yellow yolk. The deep yellow yolk is formed in the daytime. On the other hand, the light yellow yolk is formed at night when the protein concentration in blood serum is lower than in the daytime. The light yellow yolk layer (0.25-0.42 mm thick) and the deep yellow yolk layer (about 2 mm thick), thus, appear alternatively and circularly [16]. The presence of this layered structure of the egg yolk is observed when an egg from a hen that has been intravenously injected periodically with an oil-soluble pigment is hard-boiled (Figure 5).

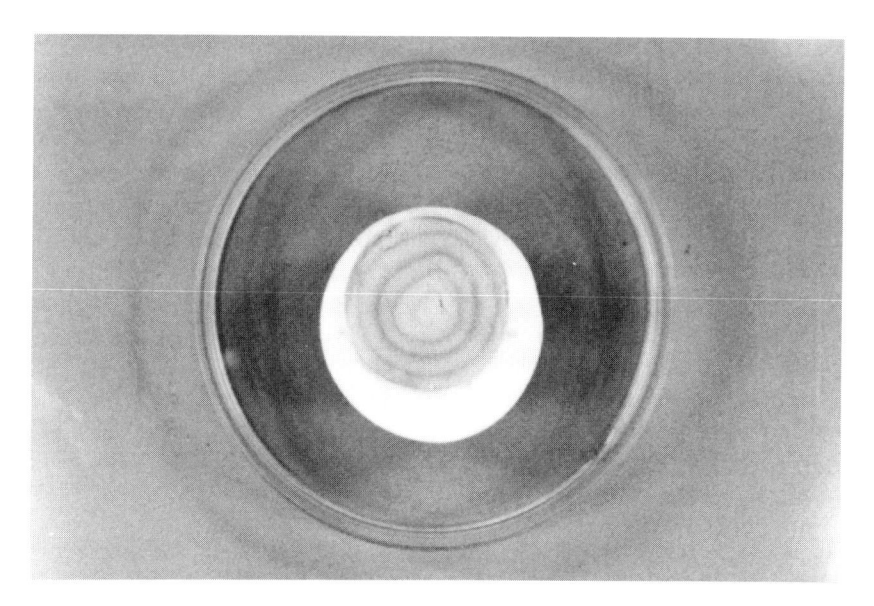

Figure 5. Structure of egg yolk. A hen was injected intravenously with 4 ml of Sudan black solution (1 mg/ml) and an egg from the hen was hard-bioled and sectioned.

III. PHYSIOLOGY OF EGG LAYING

A pair of primordiums of gonad appears during embryogenesis of the hen. In birds it is known that the right side of the gonad ceases its development around the seventh day of embryogenesis and gradually degenerates. Therefore, the ovary and oviducts of hens are those originated from the left gonad.

A. OVARY
1. Development of the ovary

As early as 1941, Chaikoff and his co-workers showed that the ovary of a newly hatchted chick weighs about 30 mg. When the chicken becomes 150 days old, the ovary has grown to about 7 g [17]. The weight of ovary rapidly increases to about 40 g (around 170 days old); then, the hen begins to lay eggs. The average weight of the ovary of egg-laying hens is 50 g. It is known that about 12,000 ova exist in a mature ovary. Each ovum becomes a follicle after being covered with a granular layer. However, most of the follicles gradually degenerate, and only about 2,000 accumulate white yolk to grow to about 6 mm in diameter. The developed follicles are called white follicles. The follicles in the ovary are surrounded well by the hen's veins [18].

2. Rapid growth of follicles in the ovary

The white follicle begins a rapid accumulation of yellow yolk in 7 to 12 days prior to ovulation. The follicles at this stage are called yellow follicles, and the rapid increase of yellow yolk terminates in 9 days on average, about a day before the matured follicle is ovulated. The weight of the follicle reaches 16-18 g at this stage. This matured follicle is ovulated at intervals of about 24-27 hr in the case of a white leghorn. Therefore, about 9 yellow follicles in a line are observed in the ovary of egg-laying hens. The rapid growth of the follicles is regulated by certain hormones (Figure 6) [19]. Both follicle stimulating hormone (FSH) and luteinizing hormone (LH) releases are necessary to induce follicular maturation and ovulation in the hen [20]. Evidently, the smaller preovulatory follicles of the domestic hen are a target tissue for FSH. During follicular maturation, the responsiveness to FSH decreases with a concomitant increase in responsiveness to LH [21]. The stages of follicular maturation of a preovulatory follicle in the hen can be divided into an extended proliferative phase (prior to LH surge) and a brief ovulatory phase (after LH surge) [22]. Larger doses of FSH increased the number of small follicles (10 mm diameter) and yolk deposition. Apparently, small follicles which have not entered the rapid growth phase are responsive to FSH [23]. Oishi et al. mentioned that a decreased egg production rate is associated with the effect of a low level of plasma LH on ovulation, the effect of a low level of plasma LH on the release or estradiol from the ovary, and the effect or a low plasma level of estradiol on the formation of precursors of egg in the liver [24]. Wells and his co-workers found that LH had the greatest stimulatory effect on the granulosa response [25].

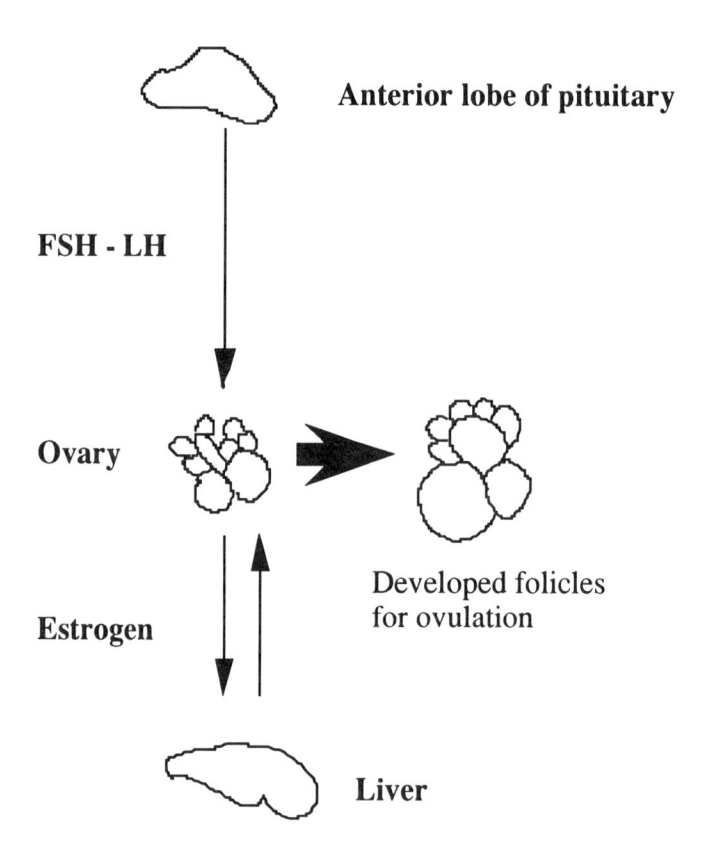

Figure 6. A proposed mechanism involved in hen ovulation. FSH, follicle stimulating hormone; LH, luteinizing hormones.
Translated from Tanaka and Koga (1988), Tasaki, Yamada, Morita, and Tanaka, Eds., [19] by permission of Yokendo, Co., Ltd., Tokyo.

B. PROCESS OF EGG LAYING

When yellow yolk is accumulated sufficiently, the follicle in the ovary is ovulated into the oviduct. Usually, it takes 24-27 hr for the encapsulated egg to come out through the oviduct. The avian oviduct is a tubular organ which extends from the ovary to the cloaca. The oviduct ensures smooth transport of the egg and secretes extracellular matrix components to surround the egg albumen. In laying hens, the oviduct is 40-80 cm long. The oviduct, with an average weight of 40 g, consists of five portions: infundibulum, albumen-secreting portion, isthmus, uterus, and vagina (Figure 7) [18].

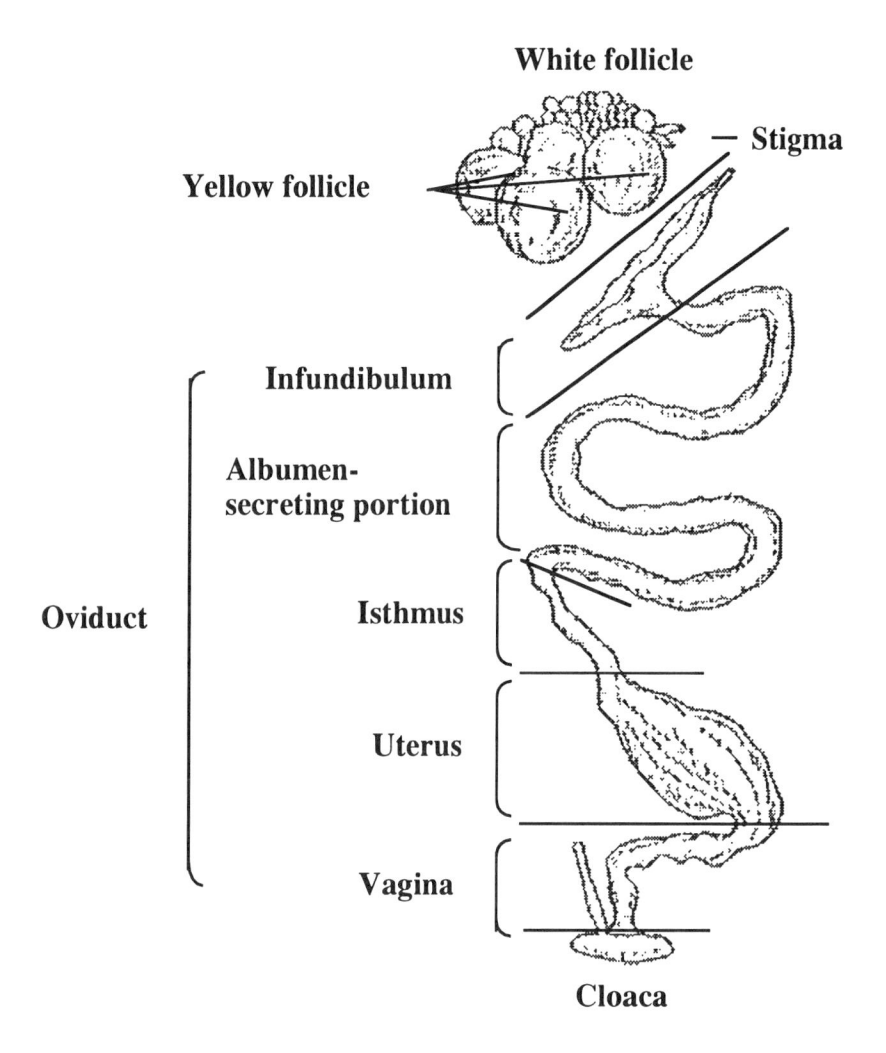

Figure 7. The follicle and parts of the oviduct.
Modified from Romanoff and Romanoff (1949) [16] by permission of John Wiley & Sons, Inc., New York.

1. Infundibulum

The infundibulum is the top portion of the oviduct and opens its ampula (broad, funnel-shaped, anterior end) towards the ovary to receive the ovulated follicles. This portion is about 11 cm long, and the ovulated follicle is held here for 15 to 30 min. During this period, the ovulated follicle has the opportunity to encounter the cock's sperm for fertilization. The infundibulum is constructed with a thick, wall-like tissue, and its mucosa consists of chorionic epithelial cells. The infundibulum has no gland, but the outer layer of the vitelline membranes of the yolk and the chalazal layer of the albumen are probably generated here [18].

2. Albumen-secreting portion

The albumen-secreting portion is about 34 cm long, the longest among the five portions of the oviduct. The follicle is held here for about 174 min. In this portion, the egg albumen is secreted to cover the egg yolk. Furthermore, it is noticeable that an amount of albumen enough for two eggs is always stored in this portion.

3. Isthmus

The isthmus is about 11 cm long. The egg yolk enveloped with albumen comes down to the isthmus, and then the shell membranes are immediately formed to wrap the egg albumen from the outer side. This process finishes in about 74 min.

4. Uterus

The uterus is only about 10 cm long. However, to complete the process of calcification, the egg is held in the uterus for about 21 hr. Several glands secrete fluid with a high concentration of calcium ion onto the inner surface of uterus. The salts in this fluid consist of sodium bicarbonate (0.5%), sodium chloride (0.24%), potassium chloride (0.16%), and calcium chloride (0.05%) [26]. The route for calcium to reach the shell is still unclear. However, the eggshell structure is formed by assembling a crystalline-like calcium structure on the egg shell membrane. In the oviposition process, the uterus also works binding arginine vasotocin (AVT) and progesterone. Takahashi and his colleagues found that the binding affinity and capacity of AVT receptor in the hen uterus changed during a period before and after oviposition. Three hours before oviposition, the binding capacity of the AVT receptor increased [27]. The same authors mention that the specific binding of the progesterone receptor in the uterus increases 2 hr before oviposition and remains high until oviposition. During oviposition, the uterine tissue produces prostaglandins which cause uterine contraction, resulting in a possible increase in the plasma AVT level [27, 28].

5. Vagina

The vagina is the last portion of the oviduct, and it is about 9 cm long. The edge of the vagina connects with the cloaca. The muscle fiber developed will serve to help in the laying of eggs. It takes only about 5 min for the egg to pass through this portion.

REFERENCES

1. **Board, R. G. and Hall, N. A.,** The cuticle: A barrier to liquid and particle penetration of the shell of the hen's egg, *Br. Poult. Sci.*, **14**, 69, 1973.
2. **Belyavin, C. G. and Boorman, K. N.,** The influence of the cuticle on eggshell strength, *Br. Poult. Sci.*, **21**, 295, 1980.
3. **Simkiss, K.,** The structure and formation of the shell and shell membranes, in *Egg Quality: A Study of the Hen's Egg,* Carter, T. C., Ed., Oliver & Boyd, Edinburgh, 1968, p 3.
4. **Cooke, A. S. and Balch, D. A.,** Studies of membrane, mammillary cores and cuticle of the hen egg shell, *Br. Poult. Sci.*, **11,** 345, 1970.

5. Baker, J. R. and Balch, D. A., A study of the organic material of hen's-egg shell, *Biochem. J.*, **82**, 352, 1962.
6. Suyama, K., Nakamura, H., Ishida, M., and Adachi, S., Lipids in the exterior structures of the hen egg, *J. Agric. Food Chem.*, **25**, 799, 1977.
7. Parsons, A. H., Structure of the eggshell, *Poult. Sci.*, **61**, 2013, 1982.
8. Candlish, J. K. and Scougall, R. K., L-5-hydroxylysine as a constituent of the shell membranes of the hen's egg, *Int. J. Protein Res.*, **1**, 299, 1969.
9. Wong, M., Hendrix, M. J. C., Mark, K. V. D., Little, C., and Stern, R., Collagen in the egg shell membranes of the hen, *Dev. Biol.*, **104**, 28, 1984.
10. Arias, J. L., Fernandez, M. S., Dennis, J. E., and Caplan, A. I., Collagens of the chicken egg shell membranes, *Connect. Tissue Res.*, **26**, 37, 1991.
11. Starcher, B. C. and King, G. S., The presence of desmosine and isodesmosine in egg shell membrane protein, *Connect. Tissue Res.*, **8**, 53, 1980.
12. Crombie, G., Sinder, R., Faris, B., and Franzblau, C., Lysine-derived cross-links in the egg shell membrane, *Biochim. Biophys. Acta*, **640**, 365, 1981.
13. Leach, R. M., Rucker, R. B., and Dyke, G. P. V., Egg shell membrane protein: A nonelastin deomosine/ isodesmosine-containing protein, *Arch. Biochem. Biophys.*, **207**, 353, 1981.
14. Bellairs, R., Harkness, M., and Harkness, R. D., The vitelline membrane of the hen's egg, *J. Ultrastructure Res.*, **8**, 339, 1963.
15. Britton, W. M., Vitelline membrane chemical composition in natural and induced yolk mottling, *Poult. Sci.*, **52**, 459, 1973.
16. Romanoff, A. L. and Romanoff, A. J., Structure, in *The Avian Egg*, John Wiley & Sons, Inc., New York, 1949, p 132.
17. Chaikoff, I. L., Lorenz, F. W., and Enternman, C., Endocrine control of the lipid metabolism of the bird. IV. Lipid metabolism of the bird during pubescence and the annual rest, *Endocrinology*, **28**, 597, 1941.
18. Burley, R. W. and Vadehra, D. V., An outline of the physiology of avian egg formation, in *The Avian Egg*, John Wiley & Sons, Inc., New York, 1989, p. 19.
19. Tanaka, K. and Koga, O., Developmental physiology, in *Yokei Handbook,* Tasaki, I., Yamada, Y., Morita, T., and Tanaka, K., Eds., Yokendo Co., Ltd., Tokyo, 1988, p 164.
20. Imai, K. and Nalbandov, A. V., Changes in FSH activity of anterior pituitary glands and of blood plasma during the laying cycle of the hen, *Endocrinology*, **88**, 1465, 1971.
21. Calvo, F. O. and Bahr, J. M., Adenylyl cyclase system of the small preovulatory follicles of the domestic hen: responsiveness to follicle-stimulating hormone and luteinizing hormone, *Biol. Reprod.*, **29**, 542, 1983.
22. Jackson, J. A., Tischkau, S. A., Zhang, P., and Bahr, J. M., Plasminogen activator production by the granulosa layer is stimulated by factor(s) produced by the theca layer and inhibited by the luteinizing hormone surge in the chicken, *Biol. Reprod.*, **50**, 812, 1994.
23. Palmer, S. S. and Bahr, J. M., Follicle stimulating hormone increases serum estradiol-17 beta concentrations, number of growing follicles and yolk deposition in aging hens (*Gallus domesticus*) with decreased egg production, *Br. Poult. Sci.*, **33**, 403, 1992.
24. Oishi, T., Yoshida, K., Moriguchi, S., and Inuzaka, S., Follicle-stimulating hormone and steroidogenesis in ovarian granulosa cells during aging in the domestic hen, *Nippon Kakin Gakkaishi (in Japanese)*, **25**, 369, 1985.
25. Wells, J. W., Walker, M. A., Culbert, J., and Gilbert, A. B., Comparison of the response *in vivo* to luteinizing hormone and follicle stimulating hormone of the granulosa of six follicles from the ovarian hierarchy in the chicken (*Gallus domesticus*), *Gen. Comp.*

Endocrinol., **59**, 369, 1985.
26. **Burmester, B. R., Scott, H. M., and Card, L. E.**, Rate of plumping of uterine eggs immersed in an artificial uterine solution, *Poult. Sci.*, **19**, 299, 1940.
27. **Takahashi, T., Kawashima, M., Kamiyoshi, M., and Tanaka, K.**, Arginine vasotocin receptor binding in the hen uterus (shell gland) before and after oviposition, *Eur. J. Endocrinol.*, **130**, 366, 1994.
28. **Murakami, Y., Fujihara, N., and Koga, O.**, Arginine vasotocin level in plasma and prostaglandin concentrations in the uterine tissue before and after premature oviposition induced by orthophosphate solution in the hen, *Nippon Kakin Gakkaishi (in Japanese)*, **28**, 11, 1991.

Chapter 2

GENERAL CHEMICAL COMPOSITION OF HEN EGGS

H. Sugino, T. Nitoda, and L.R. Juneja

TABLE OF CONTENTS

I. Introduction
II. General Chemical Composition of Hen Eggs
III. Eggshell and Eggshell Membrane
IV. Egg Yolk
    A.   Egg Yolk Proteins
            1.   Low Density Lipoprotein, LDL
            2.   High Density Lipoprotein, HDL
            3.   Phosvitin
            4.   Livetin
            5.   Other Proteins
    B.   Egg Yolk Lipids
            1.   Triglycerides
            2.   Phospholipids
            3.   Sterols
            4.   Cerebrosides
    C.   Minerals of Egg Yolk
    D.   Carbohydrates of Egg Yolk
    E.   Pigments of Egg Yolk
V. Egg White
    A.   Egg White Protein
            1.   Ovalbumin
            2.   Ovotransferrin (Conalbumin)
            3.   Ovomucoid
            4.   Ovomucin
            5.   Ovoglobulin
            6.   Lysozyme
            7.   Ovomacroglobulin
            8.   Flavoprotein
            9.   Ovoglycoprotein
            10.   Ovoinhibitor
            11.   Cystatin (Ficin-Papain Inhibitor)
            12.   Avidin
            13.   Enzymes
    B.   Lipids of Egg White
    C.   Carbohydrates of Egg White
    D.   Minerals

References

I. INTRODUCTION

An avian egg contains all the substances which are indispensable for hatching in the necessary quantity as well as in the quality. The hen eggs in the market are mostly those which are unfertilized. However, it is known that there is almost no difference in the chemical compositions between fertilized and unfertilized eggs provided that they are fresh.

In this chapter, the general chemical compositions of hen eggs in the market (egg from white Leghorn) are described with some comments about the factors affecting the egg composition.

II. GENERAL CHEMICAL COMPOSITION OF HEN EGGS

A. GENERAL COMPONENTS OF EGGS

The weight and composition of each structural part of hen eggs are more or less different depending on the species, feeding, age of hens, etc. The egg of a white Leghorn usually weighs from 50 to 63 g and the structural ratios are 9 to 11% (w/w) eggshell (shell and shell membrane), 60-63% for egg white, and 28-29% for egg yolk.

As shown in Table 1, the main chemical components of hen egg are 12% lipids, 12% proteins, and the rest (75%) is water and small amounts of carbohydrates and minerals.

Table 1

General Chemical Composition of Hen Egg (Average Values, g/egg)

	Water	Protein	Carbohydrate, Free sugar	Carbohydrate, Conjugated Oligosaccharides	Lipid	Mineral
Egg yolk	18.7	9.1	3.1	0.131	0.056	5.83	0.318
Egg white	33.0	28.9	3.5	0.132	0.165	0.002	0.231
Eggshell membrane	5.9	0.1	0.25	-	-	-	5.9
Eggshell			0.15	-	-	-	
Total	57.6	38.1	7.0	0.263	0.221	5.832	6.449

Proteins are distributed in all parts of the egg, but most of them are present in the egg yolk and the egg white amounting to 44% and 50%, respectively. The remaining 6% of protein is in the eggshell and the eggshell membrane.

Lipids in eggs include true fats, phosphorus, nitrogen and/or sugar-containing lipids, and sterols. Lipids in eggs are almost exclusively contained in the egg yolk, mainly as lipoproteins.

Carbohydrates are a minor component of hen eggs. Their average content is about 0.5 g per egg, 40% of which is present in the yolk. Carbohydrates are present as free and conjugated forms which are attached to proteins and lipids. Glucose is localized in the egg white. Mannose and galactose are present as complex carbohydrates attached to proteins.

Several pigments are contained in all parts of the egg but the chemical specificity is different. The yolk has the highest pigment content and its color and density sometimes influence the price of hen eggs.

Various minerals, which are indispensable for hatching, are contained in eggs. Most of the minerals are in conjugated form and only a small portion is present as inorganic compounds or ions (Table 2) [1].

About 94% of the minerals are in the eggshell fraction, and the rest are distributed in egg yolk and egg white.

Table 2

Inorganic Elements in Each Part of the Hen Egg [1]

Inorganic Elements	Egg Yolk (mg/egg)	Egg White (mg/egg)	Egg Shell (g/egg)
Na^+	13	53	-
Mg^{2+}	24	3	0.02
P	110	6	0.02
S	3	64	trace
Cl^-	23	51	-
K^+	21	55	-
Ca^{2+}	27	4	2.21
Fe^{3+}	2	0.3	trace
Total	223	36.3	2.25

III. EGGSHELL AND EGGSHELL MEMBRANE

Eggshell is composed of about 95% minerals, in which calcium is more than 98%. Other inorganic components include phosphorous, magnesium, and trace contents of iron and sulfur (Table 2) [1]. Major inorganic salts, carbonates and phosphates of calcium and magnesium, are in the eggshell. The calcite crystals seen in eggshell are composed of calcium carbonate, and dolomite crystals are constituted of calcium and magnesium carbonate. The dolomite structure is mechanically harder than calcite [2].

The eggshell is covered with a cuticle layer outside and the shell membrane inside. The cuticle is composed of protein, a small amount of carbohydrates such as hexosamine, galactose, mannose, fucose, and a trace amount of lipids [3]. The eggshell is also a complex compound composed of protein, polysaccharide consisting of galactosamine, glucosamine, galactose, fucose, glucose, sialic acids, and a small amount of lipids.

Eggshell membrane contains a small amount of protoporphyrin as a main pigment.

IV. EGG YOLK

A. EGG YOLK PROTEINS

Unlike egg white, egg yolk is a homogeneously emulsified fluid. The major portion of egg yolk exists as lipoproteins which are separated into plasma (supernatant on centrifugation) and granule fractions (precipitate on centrifugation). The composition of egg yolk protein and its distribution are shown in Table 3 [3-6].

1. Low density lipoprotein, LDL

LDL is the major protein accounting for 65% of the total egg yolk protein. It is characterized by its emulsifying property and this property is damaged by freezing. LDL contains about 80-89% lipids. When LDL is treated with ether, a fraction retaining at the interface between ether and water appears which is called lipovitellenin. Lipovitellenin contains about 40% lipids and is soluble in NaCl saturated ether. Lipovitellenin is, thus, a derived lipoprotein from LDL by denaturation with ether.

Table 3

Egg Yolk Proteins and Their Distribution

Fraction	Plasma	Granule	Molecular weight (x 10^4)
Protein	72 % (100)	22 % (100)	
High density lipoprotein (HDL) α-Lipovitellin β-Lipovitellin	- -	(41) (29)	40 40
Low density lipoprotein (LDL)	(87)	(12)	480
Phosvitin	-	(17)	3.6
Livetin	(13)	-	α : 8.0 β : 4.2 γ : 18

( ) indicates weight distribution , %.
Modified from Nishimoto and Asano (1987), Asano and Ishihara, Eds., [3] by permission of Korin Co., Ltd., Tokyo, 1987.

2. High density lipoprotein, HDL

HDL consists of α- and β-lipovitellins. Both of them contain 21-22% lipids [1]. HDL exists as a complex with a phosphoprotein, "phosvitin."

3. Phosvitin

Phosvitin is a phosphoprotein containing about 10% phosphorus. About 80% of the phosphorus found in the egg yolk is localized in phosvitin [3]. It is very characteristic that phosvitin contains 54% of serine, whereas no content of methionine, tryptophan or tyrosine are observed. The serine residues of phosvitin are exclusively present as esters of phosphoric acid. Under low ionic strength and acidic conditions, phosvitin becomes water soluble and available in complexing with Ca^{2+}, Mg^{2+}, Mn^{2+}, Co^{2+}, Fe^{2+}, and Fe^{3+}. This is the reason that phosvitin acts as a carrier of Ca^{2+} or Fe^{2+}.

4. Livetin

Livetin is a water-soluble protein which accounts for 30% of the plasma proteins. Most of the enzymes in eggs are observed in this fraction. The enzymes observed are α–amylase, cholinesterase, phosphatase, etc., though their activities are not necessarily large. The livetin fraction is known to stand for plasma proteins in the blood of mammals immunologically, α–livetin is for serum albumin, β– and γ–livetin, for $α_2$-glycoprotein, and γ–globulin, respectively [7].

5. Other proteins

A riboflavin-binding protein exists in the egg yolk [5]. It is a hydrophilic phosphoglycoprotein conjugating one mole of riboflavin per mole of apoprotein with a molecular weight of 3.6×10^6.

B. EGG YOLK LIPIDS

Lipids are the main components of the egg yolk, and they occupy about 60% of the egg yolk based on dry weight. The lipids include triglycerides, phospholipids, cholesterol, cerebroside, and some other minor lipids. The major fatty acids are oleic, palmitic, linoleic, and stearic acids [1].

1. Triglycerides

The major fatty acids of triglycerides are oleic acid (18:1), and palmitic acid (16:0). Linoleic acid (18:2) and stearic acid (18:0) are also significant [1]. Fatty acids like myristic acid (14:0) and others are contained in trace amounts.

2. Phospholipids

Phospholipids are an important component of egg yolk lipoproteins. They contain both hydrophilic head groups and lipophilic fatty acid groups. Therefore, the phospholipids show emulsifying properties. Egg yolk is an excellent natural emulsifier and it is used widely in the food industry and in home cooking (See Chapter 8). The major components of egg yolk phospholipids are phosphatidylcholine (PC) and phosphatidylethanolamine (PE). Lysophosphatidylcholine (LPC) and lysophosphatidylethanolamine (LPE) are also present [8-11]. There are two types of phospholipids, α and β type. They differ from each other in the position of the phosphoryl ester in the glycerol skeleton of the two fatty acids groups bound to the phospholipids. The fatty acids linked at the β-position are mostly unsaturated fatty acids, whereas those linked at α- position are usually saturated fatty acids. It is a very important feature that various type of fatty acids are contained in the egg yolk phospholipids. Polyunsaturated fatty acids (PUFA) are especially concentrated in the phospholipid fraction. Almost all arachidonic acid (20:4) and docosahexaenoic acid (DHA) (22:6), which have various bioactivities, are also concentrated in the phospholipid fraction. This may be so because lysophospholipid acyltransferase contributes to the biosynthesis of phospholipids for animals living on the land and has the selectivity of PUFA as its substrates [11].

Egg yolk phospholipids, such as egg yolk lecithin, have been utilized in the food, cosmetics, pharmaceutical, and medical fields. This will be discussed more detail in Chapter 6 on egg yolk lipids.

3. Sterols

Almost all the sterol of egg yolk is cholesterol. Cholesterol is present at about 1.6% in raw egg yolk and about 5.0% in egg yolk lipids. Free cholesterol is about 84% of the total cholesterol and, the remaining 16% is cholesterol ester [12]. Most of the egg yolk cholesterol originates from the feed and some of it is synthesized when egg yolk is formed inside the hen's body.

4. Cerebrosides

Cerebrosides are classified as glycolipids. They are composed of sugar (galactose, sucrose), sphingosine, and a nitrogen-containing base. Two types of cerebrosides, ovophrenosin and ovokerasin, have been separated from egg yolk [13]. Their chirality is very similar except for the fatty acid content. Their chirality is also different; ovophrenosin is dextrorotatory whereas ovokerasin is levorotatory.

C. MINERALS OF EGG YOLK

Egg yolk contains 1% of minerals, and phosphorus is the most abundant mineral component. More than 61% of the total phosphorus of egg yolk is contained in phospholipids. Table 2 shows the content of the major minerals in egg yolk [1].

D. CARBOHYDRATES OF EGG YOLK

The content of carbohydrates in egg yolk is about 1.0%. 0.7% of it are oligosaccharides bound to protein, and they are composed of mannose and glucosamine. The remaining 0.3% is free carbohydrates in the form of glucose.

E. PIGMENTS OF EGG YOLK

Egg yolk is the portion of the egg richest in pigments in hen egg, but the quantity is 0.02% based on the dry weight. The egg yolk pigments are carotenes (classified as lypochromes) and riboflavin (classified as lyochromes) [14]. Carotenes, which are the cause of the color of the yolk, cannot be synthesized by the hen's metabolism. The hen's feed is the major factor affecting carotenes content and the color of the egg yolk.

Egg yolk carotenes are classified as xanthophils, which have an OH group in their molecules, and carotenes, which do not have it. Lutein, zeaxanthin, and cryptoxanthin belong to the xanthophil group, and β–carotene belongs to the carotene group. The quantities of each carotene in egg yolk are shown in Table 4 [1].

Table 4

Carotenoid Composition of Egg Yolk

Pigments	Content (%)
Carotene	
α - carotene	trace
β - carotene	0.03
Xanthophil	
Cryptoxanthin	0.03
Lutein	0.1
Zeaxanthin	0.2

Modified from Romanoff and Romanoff (1949) [1] by permission of John Wiley & Sons, Inc., New York.

V. EGG WHITE

Egg white occupies about 60% of the whole egg by weight. Water and proteins are the major components, accounting for about 88% and 10.4%, respectively [16]. Carbohydrates, minerals, and lipids are minor components (Table 1). Structurally, egg white is composed of four concentric layers: external thin albumen, internal thin albumen, thick albumen, and chalaziferous layer. The content of each layer is about 16.8%, 57.3%, 23.2%, and 2.7%, respectively [16]. Although the water and protein compositions are the same, except for the level of ovomucin, the thick albumen is characterized by its viscosity, which is much higher than those of external and internal thin albumens [15]. The difference in the viscosity between those layers is considered to be due to the different levels of ovomucin content.

A. EGG WHITE PROTEINS

Egg white usually contains about 11% proteins which consists of more than 40 different kinds of proteins. Many of them are still uncharacterized because of their low concentration. Ovalbumin is the major protein and constitutes about 54 % of the total egg white protein [16]. Ovotransferrin (conalbumin) and ovomucoid occupy about 12% and 11%, respectively (Table 5) [16]. Other proteins are ovomucin, ovoglobulin, lysozyme, ovomacroglobulin, ovoglycoprotein, flavoprotein, ovoinhibitor, Cystatin (ficin-papain inhibitor), and avidin.

Table 5

Composition and Physical Properties of Egg White Proteins [16]

	Content % of total protein	Isoelectric point	Molecular weight ($\times 10^4$)
Ovalbumin	54	4.5	4.5
Ovotransferrin	12	6.1	7.6
Ovomucoid	11	4.1	2.8
Ovomucin	3.5	4.5-5.0	-
Ovoglobulin G2	4	5.5	4.9
Ovoglobulin G3	4	5.8	4.9
Lysozyme	3.4	10.7	1.4
Ovomacroglobulin	0.5	4.5	-
Ovoglycoprotein	1	3.9	2.4
Flavoprotein	0.8	4	3.2
Ovoinhibitor	1.5	5.1	4.9
Cystatin (Ficin-papain inhibitor)	0.05	~5.1	1.3
Avidin	0.5	10	6.8

Modified from Burley and Vadehra (1989) [16] by permission of John Wiley & Sons, New York.

1. Ovalbumin

As mentioned before, ovalbumin is the major protein of egg white. Ovalbumin is a single peptide chain molecule with a carbohydrate side chain. Its molecular weight has been estimated to be about 4.5×10^4. Based on its phosphate residue level, ovalbumin is classified into three molecular species: Ovalbumin A1, A2, and A3. Among egg white proteins, ovalbumin is the only protein which has free SH groups. On denaturation of ovalbumin, three sluggish SH groups and one unreactive SH group turn into four chemically reactive SH groups [17]. The carbohydrate moiety of ovalbumin is consisted mainly of mannose and N-acetylglucosamine linked with asparagine residue to form N-glycosidic linkage. During preservation of hen eggs, ovalbumin turns into a more stable form of ovalbumin (named s-ovalbumin) because raising the pH increases the thermal stability. Ovalbumin greatly affects the physical nature of egg white in the nature of its coagulating and foaming properties.

2. Ovotransferrin (Conalbumin)

Ovotransferrin is similar to serum transferrin in animals. They are different only in the carbohydrate chain moiety. Ovotransferrin is a single polypeptide, and its molecular weight is about 7.6×10^4. Ovotransferrin has the ability to bind various metal ions in the ratio of two moles of ion per mole of protein. The binding affinity of the protein to metal ions is as follows: $Fe^{2+} > Cu^{2+} > Zn^{2+}$ [18]. The amino acid residues of ovotransferrin, including three tyrosine, two histidine, and one tryptophan, are reported to participate in forming complexes with metal ions. Metal complexes formed by ovotransferrin occasionally develop colors depending on the bound metal ions, reddish color with Fe^{2+}, yellow color with Cu^{2+}. In forming metal complexes, the physicochemical properties of ovotransferrin change. For example, there is a rise of pasteurization temperature and an increase in resistance to proteolysis. Also, formation of an ion complex by ovotransferrin inhibits the growth of microorganisms which require iron. This bacteriostatic activity may be one of most important biological properties of ovotransferrin. Temperature directly affects the functional properties of ovotransferrin, therefore, aluminum salt is added, if necessary, to egg white before thermal sterilization, to prevent coagulation and loss of foaming properties [18].

3. Ovomucoid

Ovomucoid is a glycoprotein which has trypsin inhibitor activity with a high thermal stability. Its molecular weight is about 2.8×10^4. Ovomucoid is consisted of 3 domains: I, II, and III. The active site of the trypsin inhibitor is located in domain II. It has been revealed that the secondary structure of ovomucoid consists of 26 % α–helix, 46 % β–configuration, 10 % β-turn, and 18 % random coil [19]. Each domain shows a good recovery of tertiary structure after denaturation by denaturing agents [20]. The reversibility of ovomucoid denatured by heat is thought to be by virtue of its tertiary structure. Ovomucoid has nine disulfide bonds in its molecule, three for each domain. These disulfide bonds do not contribute to linkage between domains, but they serve for trypsin inhibition activity. Ovomucoid is a Kazal-type inhibitor, and, in eggs of most birds, its active site is the lysine residue. However, in chicken eggs, the active site is arginine [21]. Ovomucoid contains 20-25% carbohydrates. The carbohydrates consist of N-acetylglucosamine (12.5-15.4%), mannose (4.3-4.7%), galactose (1.0-1.5%), and N-acetylneuraminic acid (0.4-4.0%). The carbohydrates chains link with asparagine residue, forming N-glycoside linkage with N-acetylglucosamine. Ovomucoid splits into at least five bands on an electropheretogram due to the different levels of N-acetylneuraminic acid.

4. Ovomucin

As described above, the specific jellying property of egg white is attributed to the ovomucin. Ovomucin is a glycoprotein with an extremely large molecular weight. It is made up a soluble and an insoluble fraction. Both of them consist of two associated subunits, α–ovomucin and β–ovomucin. Soluble ovomucin consists of forty α–ovomucin molecules and three β–ovomucin molecules (molecular weight is about 8.3×10^6) [22]. Insoluble ovomucin consists of 84 α–ovomucin molecules and 20 β–ovomucin molecules (molecular weight is about 2.3 x 10^7) [22]. Structurally, insoluble ovomucin is the major component of the gel-like insoluble fraction of the thick albumen of egg white, whereas, soluble ovomucin is the main component of the outer and inner thin albumen. Egg white β–ovomucin contains far more saccharide chains than α–ovomucin. α–Ovomucin contains hexoses (6-7%), hexosamine (7-8%), N-acetylneuraminic acid (0.3-0.8%), and sulfate (0.2 -0.4%), whereas β–ovomucin contains hexose (18-20%), hexosamine (18-22%), N-acetylneuraminic acid (11-15%), and sulfate (3%) [23]. Ovomucin is involved in supporting the density, and keeping the nature of egg white. Ovomucin also inhibits hemagglutination by viruses [24]. During storage, the thick albumen of egg white changes into the thin albumen being due to a change in ovomucin.

5. Ovoglobulin

In early studies, six globulin fractions were thought to be present in egg white. They are macroglobulin, ovoglobulin G1, G2, G3, and two other globulins. However, it was found later that the two globulins were ovoinhibitors, and ovoglobulin G1 was identified as lysozyme. Presently, the name of ovoglobulin is given only to ovoglobulins G2, and G3. Their molecular weights are both 3.6-4.5 x 10^4. The biological function of ovoglobulin has not been made clear, but ovoglobulin shows an important role in the foaming property of egg white.

6. Lysozyme

Lysozyme was first discovered in nasal mucus by Fleming in 1922 [25]. Soon, the same bacteriolytic action was discovered in egg white. The name lysozyme was originated from the enzyme which has lytic action against bacterial cells. Lysozyme of egg white consists of 129 amino acid residues with a molecular weight of 14,400. Because of its basic character, lysozyme binds to ovomucin [26], transferrin, [27], or ovalbumin [28] in egg white. The chalaziferous layer and chalaza bind lysozyme in 2-3 times more than other egg white proteins [29]. Lysozyme is highly stable in acidic solution and keeps its activity even after 1-2 minutes heating at 100°C. One of the reasons for the thermal stability of lysozyme is thought to be certain two of four disulfide bonds contained. Lysozyme catalyzes the hydrolysis of the β 1:4-glycosidic linkage between N-acetylmuraminic acid and N-acetylglucosamine in polysaccharide, which is a major component of certain bacterial cell walls. It also catalyzes the hydrolysis of the β 1:4-glycosidic linkage between N-acetylglucosamines in chitin, a polymer of N-acetylglucosamines [30].

7. Ovomacroglobulin

Ovomacroglobulin is the second largest glycoprotein next to ovomucin, and its molecular weight is 7.6-9.0 x10^5. Ovomacroglobulin, in the same way as ovomucin, has the ability to inhibit hemagglutination.

8. Flavoprotein

After being transported from the blood to the egg white, most of the riboflavin (vitamin B$_2$) is stored in the egg white binding with the apoprotein. This is flavoprotein. The apoprotein is acidic with a molecular weight of 3.2-3.6 x 10^4. The apoprotein contains a carbohydrate moiety (14%) made up of mannose, galactose, and glucosamines, 7-8 phosphate groups, and 8 disulfide bonds. One mole of apoprotein binds one mole of riboflavin, but this binding ability is lost at pH lower than 4.2, which is its isoelectric pH.

9. Ovoglycoprotein

Ovoglycoprotein is an acidic glycoprotein. Its molecular weight is 24,400. This protein contains hexoses (13.6%), glucosamine (13.8%), and *N*-acetylneuraminic acid (3%) [31]. The biological functions of ovoglycoprotein are still unclear.

10. Ovoinhibitor

Matsushima found a trypsin inhibitor other than ovomucoid in egg white. This trypsin inhibitor was named ovoinhibitor [32]. Ovoinhibitor is a Kazal-type inhibitor as ovomucoid, but, ovoinhibitor functions as a multiheaded inhibitor and inhibits bacterial serine proteinase, fungal serine proteinase, and chymotrypsin in chicken.

11. Cystatin (Ficin-Papain inhibitor)

The third proteinase inhibitor in egg white is cystatin (ficin-papain inhibitor), which inhibits ficin and papain. In contrast to ovomucoid, ficin inhibitor is a small molecule (molecular weight 12,700), has no carbohydrates, and shows a high thermal stability [33].

12. Avidin

Avidin is a basic glycoprotein composed of four subunits (molecular weight is 15,600 each) made up of 128 amino acid residues. This protein is a trace component (0.05%) of egg white, but it has been well studied because of its ability to bind to biotin [34], one of vitamin B group. By binding biotin, avidin makes the vitamine inactive. Therefore, avidin is considered to induce deficiency of vitamin B. However, the amount of avidin is so small in an ordinary diet that its negative effect is not considered important. Each subunit of avidin binds to a molecule of biotin [35]. The binding between biotin and avidin is so strong that heating the complex at 120°C for 15 min is necessary to separate them.

13. Enzymes

Albumen is reported to contain, some other enzymes than lysozyme. They are phosphatase, catalase, glycosidase, etc., which are mentioned in Chapter 9.

B. LIPIDS OF EGG WHITE

Fresh egg white contains only a trace amount of lipid (about 0.02%) [36]. However, in eggs stored for a long time when the yolk membrane becomes weak, triglycerides and cholesterol esters are considered to move into the albumen, causing a change in the foaming nature of egg white [36].

C. CARBOHYDRATES OF EGG WHITE

Carbohydrates of egg white are present in both free form (0.4% of egg white) and binding form (0.5% of egg white) as glycoprotein. Most free form is glucose (98% of free form). Small amounts of fructose, mannose, arabinose, xylose, and ribose have also been detected [37, 38]. Since they are reducing sugars, they participate in aminocarbonyl reactions which causes browning of powder products of whole egg or egg white.

D. MINERALS

The major inorganic components of egg white are sulfur, potassium, sodium, and chlorine. Phosphorus, calcium, and magnesium are next in importance. Iron is a trace component. Table 2 shows the content of the major minerals in egg white.

REFERENCES

1. **Romanoff, A. L. and Romanoff, A. J.,** Biophysicochemical constitution. Chemical composition, in *The Avian Egg*, Romanoff, A. L. and Romanoff, A. J., Eds., John Wiley & Sons, New York, 1949, p 311.
2. **Sato, Y.,** Keiran no Kagaku, in *Shokuran no Kagaku to Riyo (in Japanese)*, Sato, Y., Ed., Chikyusha, Tokyo, 1980, p 45.
3. **Nishimoto, K. and Asano, Y.,** Egg chemistry, in *Egg- Its Chemistry and Processing Technology (in Japanese)*, Asano, Y and Ishihara, R., Eds., Korin Co., Ltd., Tokyo, 1987, p 51.
4. **Nakanishi, T., Hujimaki, M., Andou, N., Sato, Y., and Nakamura, R.,** *Kaitei Shinban Chikusanriyougaku (in Japanese)*, Asakurasyobou, 1946.
5. **Osuga, D. T. and Feenry, R. E.,** Egg proteins, in *Food Proteins*, Whitaker, J. R. and Tannenbaum, S. R., Eds., AVI. Pub. Co., Westport, Connecticut, 1977, p 209.
6. **Vadehra, D. V. and Nath, K. R.,** Eggs as a source of protein, *CRC Crit. Rev. Food Tech.*, 4, 193, 1973.
7. **Williams, J.,** Serum proteins and the livetins of hen's egg yolk, *Biochem. J.*, 83, 346, 1962.
8. **Riemenschneider, R. W., Ellis, N. R., and Titus, H. W.,** The fat acids in the lecithin and glyceride fractions of egg yolk, *J. Biol. Chem.*, 126, 225, 1938.
9. **Levene, P. A. and West, C. J.,** Cephalin. III. Cephalin of the egg yolk, kidney, and liver, *J. Biol. Chem.*, 24, 111, 1916.
10. **Levene, P. A.,** Sphingomyelin. III., *J. Biol. Chem.*, 24, 69, 1916.
11. **Rhodes, D. N.,** The effect of cod-liver oil in the diet on the composition of hen's egg phospholipids, *Biochem. J.*, 68, 380, 1958.
12. **Miyamori, S.,** Determination of cholesterol by Embden method, *Nagoya, J. Med. Sci.*, 8, 135, 1934.
13. **Levene, P. A. and West, C. J.,** Cerebrosides. V. Cerebrosides of the kidney, liver, and egg yolk, *J. Biol. Chem.*, 31, 649, 1917.
14. **Karrer, P. and Schöpp, K.,** 77. Isolation of lyochrome from egg yolk (Ovoflavin g), *Helv. Chim. Acta.*, 17, 735, 1934.
15. **Sato, Y. and Hayakawa, S.,** Further inspection for the structure of thick egg white, *Nippon Nogeikagaku Kaishi*, 51, 47, 1977.
16. **Burley, R. W. and Vadehra, D. V.,** The albumen: Chemistry, in *The Avian Egg- Chemistry and Biology*, Burley, R. W. and Vadehra, D. V., Eds., John Wiley & Sons, New York, 1989, p 65.
17. **Fernandez-Diez, M. J., Osuga, D. T., and Feeney, R. E.,** The sulfhydryls of avian ovalbumins, bovine β-lactoglobulin, and bovine serum albumin, *Arch. Biochem. Biophys.*, 107, 449, 1964.
18. **Donovan, J. W. and Ross, K. D.,** Nonequivalence of the metal binding sites of conalbumin, *J. Biol. Chem.*, 250, 6022, 1975.
19. **Watanabe, K., Masuda, T., and Sato, Y.,** The secondary structure of ovomucoid and its domains as studied by circular dichroism, *Biochim. Biophys. Acta*, 667, 242, 1981.
20. **Matsuda, T., Watanabe, K., and Sato, Y.,** Secondary structure prediction of chicken egg white ovomucoid, *Agric. Biol. Chem.*, 45, 417, 1981.
21. **Lin, Y. and Feeney, R. E.,** Ovomucoids and ovomucininhibitors, in *Glycoproteins*, Gottschalk, A., Ed., Elsevier, Amsterdam, 1972, p 762.
22. **Hayakawa, S. and Sato, Y.,** Subunit structures of sonicated α and β-ovomucin and their molecular weights estimated by sedimentation equilibrium, *Agric. Biol. Chem.*, 42, 957,

1978.
23. **Kato, A. and Sato, Y.,** The separation and characterization of carbohydrate rich component from ovomucin in chicken eggs, *Agric. Biol. Chem.*, **35**, 439, 1971.
24. **Sugihara, T. F., Macdonnell, L. R., Knight, C. A., and Reeney, R. E.,** Virus antihemagglutinin activities of avian egg components, *Biochim. Biophys. Acta*, **16**, 404, 1955.
25. **Fleming, A.,** On a remarkable bacteriolytic element found in tissues and secretions, *Proc.. R. Soc. London, Ser B.*, **93**, 306, 1922.
26. **Kato, A., Imoto, T., and Yagishita, K.,** The binding groups in ovomucin-lysozyme interaction, *Agric. Biol. Chem.*, **39**, 541, 1975.
27. **Ehrenpreis, S. and Warner, R. C.,** The interaction of conalbumin and lysozyme, *Arch. Biochem. Biophys.*, **61**, 38, 1956.
28. **Nakai, S. and Kason, C. M.,** A fluorescence study of the interactions between κ- and α_{s1}-casein and between lysozyme and ovalbumin, *Biochim. Biophys. Acta*, **351**, 21, 1974.
29. **Baker, R. C., Hartsell, S. E., and Stadelman, W. J.,** Lysozyme studies on chicken egg chalazae, *Food Res.*, **24**, 529, 1959.
30. **Berger, L. R. and Weisler, R. S.,** The β-glucosaminidase activity of egg-white lysozyme, *Biochim. Biophys. Acta*, **26**, 517, 1957.
31. **Ketterer, B.,** Ovoglycoprotein, a protein of hen's egg white, *Biochem. J.*, **96**, 372, 1965.
32. **Matsushima, K.,** An undescribed trypsin inhibitor in egg white, *Science*, **127**, 1178, 1958.
33. **Sen, L. C. and Whitaker, J. R.,** Some properties of a ficin-papain inhibitor from avidin egg white, *Arch. Biochem. Biophys.*, **158**, 623, 1973.
34. **Eakin, R. E., Snell, E. E., and Williams, R. J.,** The concentration and assay of avidin, the injury-producing protein in raw egg white, *J. Biol. Chem.*, **140**, 535, 1941.
35. **Green, N. M. and Toms, E. J.,** The properties of subunits of avidin coupled to sepharose, *Biochem. J.*, **133**, 687, 1973.
36. **Sato, Y., Watanabe, K., and Takahashi, T.,** Lipids in egg white, *Poult. Sci.*, **52**, 1564, 1973.
37. **Tunmann, P. and Silberzahn, H.,** The carbohydrate in hen egg, *Z. Lebensm. Unters. Band.*, **115**, 121, 1961.
38. **Nonami, Y.,** Non-microbial deterioration of preserved egg albumen during the storage. part IV. dialyzable sugars of preserved egg, *Nippon Nogeikagaku Kaishi (in Japanese)*, **33**, 997, 1959.

Chapter 3

NUTRITIVE EVALUATION OF HEN EGGS

M.A. Gutierrez, H. Takahashi, and L.R. Juneja

TABLE OF CONTENTS

I. Introduction
II. Nutritional Value of Hen Eggs
    A. Protein
    B. Lipids
    C. Carbohydrates
    D. Vitamins
    E. Minerals
III. Factors Affecting the Composition of Hen Eggs
IV. Some Drawbacks of Eating Eggs
    A. Allergy Induced by Egg Protein
    B. Bioavailability of Vitamins and Iron
    C. Cholesterol of Hen Egg
V. Final Remarks

References

I. INTRODUCTION

Because of their incomparably nutritious properties, avian eggs have been highly esteemed as a wholesome food for centuries. In addition to the nutritional value, they have played a special role in human daily livelihood due to their easy digestibility, good taste, and numerous applications in preparing a wide variety of foods. The avian egg is an encapsulated material indispensable for the development of the embryo. The chicken embryo develops into a chick by utilizing only the materials inside the shell at an adequate temperature. The protein, lipids, carbohydrate, vitamins, and minerals contained in the egg are all necessary and sufficient for developing the chicken's body. The high nutritional properties of eggs make them ideal for many people with special dietary requirements. Eggs are also suitable for nutritional enhancement of several kinds of foods. On top of everything, eggs are safe, familiar, and economically available in large quantities.

II. NUTRITIONAL VALUE OF HEN EGGS

Hen eggs contain 74.57% water. However, they have four major nutritional components: proteins (12.14%), lipids (11.15%), all necessary vitamins (except vitamin C), and minerals [1]. The nutritional value of egg proteins has been extensively evaluated. Eggs are classified in the protein food group with meat, poultry, and fish. Egg proteins consist of an ideal balance of nutritionally indispensable amino acids. The hen egg is, thus, useful as a supplement to those foods in which essential amino acids are in low concentration or absent. Eggs are also an excellent source of essential fatty acids. Lipids contained in the egg yolk provide most of the metabolic energy necessary for the development of the embryo. The hen egg is also a useful food as a source of minerals and vitamins. In spite of the present technology in processed baby foods, egg yolk is still a practical and suitable nutritional supplement for young babies.

High nutritional value, low caloric content, blandness, and ease of digestibility are characteristics that make eggs ideal for young or old people, healthy or convalescent.

A. PROTEIN

Egg proteins, which are distributed in both yolk and albumen, are nutritionally complete proteins with an unsurpassable balance of amino acids. The amino acid composition of hen eggs is shown in Table 1. The protein value of whole egg proteins is considered to be 100. According to the World Health Organization, egg protein has the highest true digestibility among major food proteins. Because of its high quality, egg protein is used as a standard for measuring the nutritional quality of other food proteins. Table 2 shows the suggested daily requirements of essential amino acids and the essential amino acid composition of casein, soybean, and egg proteins [2]. The protein content of two eggs is about 12 g, which corresponds to 30% of the dietary allowance recommended by the National Research Council (1980) in the United States. [3]. It has been reported that the body weight gain of rats as a function of food efficiency and the protein efficacy ratio (PER) is better on an egg protein diet than that on a lactoalbumin or casein diet [4]. Egg white protein digested with proteolytic enzymes is used as a liquid protein diet for clinical purposes.

Table 1
Amino acid composition of Hen Eggs (g per egg) [1]

	Yolk	Albumen
Alanine	0.140	0.215
Arginine	0.193	0.195
Aspartic Acid	0.233	0.296
Cystine	0.050	0.083
Glutamic Acid	0.341	0.467
Glycine	0.084	0.125
Histidine	0.067	0.076
Isoleucine	0.160	0.204
Leucine	0.237	0.291
Lysine	0.200	0.250
Methionine	0.171	0.130
Phenylalanine	0.121	0.210
Proline	0.116	0.126
Serine	0.231	0.247
Threonine	0.151	0.149
Tryptophan	0.041	0.051
Tyrosine	0.120	0.134
Valine	0.170	0.251

USDA 1976 [1].

Table 2
Suggested Essential Amino Acid Daily Requirements (mg/day) Compared with the Composition of Several Proteins (mg)

	Suggested Daily Requirement (mg/day) [a]				Egg protein	Casein	Soybean protein
	Infant mean	Preschool child (2-5 years)	School child (10-12 years)	Adult			
Histidine	26	19	19	16	25	31	29
Isoleucine	46	28	28	13	55	57	53
Leucine	93	66	4	19	89	97	86
Lysine	66	58	44	16	72	82	67
Methionine+cystine	42	25	22	17	59	35	28
Phenylalanine+tyrosine	72	63	22	19	93	110	99
Threonine	43	34	28	9	46	43	38
Tryptophan	17	11	9	5	15	13	14
Valine	55	35	25	13	67	70	53
Total	460	339	241	127	521	538	467

[a]FAO/WHO/UNU [2].

B. LIPIDS

Almost egg lipids are contained in the yolk. Egg yolk is considered a potentially important source of energy because more than 65% of the contents of dry yolk is lipids. Egg yolk contains triglycerides, phospholipids, and sterols. The composition of fatty acids constituting triglycerides and phospholipids is described in detail in Chapter 6. Hen eggs are a rich source of linoleic acid, which is essential in human nutrition. The fatty acids in eggs are more unsaturated than those of most animal lipids (Table 3). The recognition of the value of egg yolk phospholipids in the human diet is being increasingly appreciated. Egg yolk phosphatidylcholine is a significant source of choline, which is an important nutrient in brain development, liver function, and cancer prevention [5, 6]. When hens are fed fish oil or flax seeds, the long-chain omega-3 polyunsaturated fatty acids (ω-3 PUFA) are incorporated in the sn-2 position of yolk phospholipids [7, 8]. This fact makes egg yolk from those hens an ideal supplement for various foods, mainly infant formula [9-11].

ω-3 PUFA are now regarded as essential in the diet for brain function and visual acuity in humans. They are especially important for pregnant and nursing mothers and their infants. It was suggested that ω-3 PUFAs might be indispensable for neural development of newborn children [12]. Infants usually receive ω-3 PUFAs from breast milk. However, in these modern times, mothers eat modern diets, which usually are low in this essential fatty acid [13]. ω-3 PUFA, can be biologically incorporated into egg yolk lipids by manipulating the chicken feed [14]. Therefore, the chicken egg is a potential vehicle to provide the necessary ω-3 PUFA.

Table 3
Lipid Composition of Hen Eggs [1]

Lipids	Whole (g)	Yolk (g)	Albumen
Fatty acids			
Saturated	1.67	1.68	0
14:0	0.02	0.02	0
16:0	1.23	1.24	0
18:0	0.43	0.43	0
Monounsaturated	2.23	2.24	0
14:1	0.005	0.005	0
16:1	0.19	0.19	0
18:1	2.04	2.05	0
Polyunsaturated	0.72	0.73	0
18:2	0.62	0.62	0
18:3	0.02	0.02	0
20:4	0.05	0.05	0
Cholesterol	0.264	0.258	0
Lecithin (Phosphatidylcholine)	1.27	1.22	0
Cephalin (Phosphatidylethanolamine)	0.253	0.241	0

USDA 1976 [1].

C. CARBOHYDRATES

Most of the egg carbohydrates are in the albumen. The content of carbohydrates in hen egg is only about 1.0% of the whole egg on a wet basis. Thus, the carbohydrates of hen egg cannot be considered a source of energy. Recently, the structures and physiological function of the carbohydrates in egg have been extensively studied. It was found that egg yolk contains oligosaccharides which have *N*-acetylneuraminic acids at the nonreducing end. It has been suggested that sialyloligosaccharides play important roles, such as prevention of viral and bacterial infection and maintenance of normal cell growth [15, 16]. The carbohydrates of hen egg will be described in detail in Chapter 7 of this book.

D. VITAMINS

Most egg vitamins, especially the fat soluble vitamins, are contained in the yolk. Hen egg is considered a source of most vitamins necessary for human nutrition, except vitamin C. Table 4 shows the vitamin content of one egg [17]. One egg may supply almost 12 % vitamin A, more than 6% of vitamin D, 9% riboflavin, and 8% panthotenic acid of the recommended daily allowance in the United States (Table 4). Only fish liver oils contain more vitamin D than eggs.

Antitumor activity of carotenoids and retinoids is one current topic. These compounds seem to play an important role in scavenging several peroxide radicals. The modified eggs enriched in α-tocopherol, β-carotene and retinol which are obtained by supplementing these substances in hens' feed, are highly regarded nutritionally [18].

Table 4
Vitamins in Hen Egg Compared with the Recomended Daily Allowance

	RDA (USA)[a]	RDA (Japan)[b]	Amount in an egg[c]	c/a x 100 (%)
Vitamin A	5000 IU	2000 IU	590 IU	11.8
Vitamin D	400 IU	100 IU	25 IU	6.3
Vitamin E	30 IU	8 mg[d]	1 IU	3.3
Vitamin C	60 mg	50 mg	-	-
Folic acid	0.4 mg	-	2.5 mg	0.625
Thiamine	1.5 mg	1.0 mg	0.055 mg	3.7
Riboflavin	1.7 mg	1.4 mg	0.15 mg	8.8
Niacin	20.0 mg	17 mg	0.05 mg	0.3
Vitamin B6	2.0 mg	-	0.13 mg	6.5
Vitamin B12	6.0 mg	-	0.14 mg	2.3
Biotin	0.3 mg	-	10 mg	3.7
Pantothenic acid	10.0 mg	-	0.81 mg	8

[a] Recommended daily Allowances in the United States [3].

[b] Recommended daily Allowances for the Japanese (1994) [20].

[c] American egg board [17].

[d] α-tocopherol equivalent

E. MINERALS

Minerals in eggs usually depend on the kind of feed, age of hen, environment conditions, etc. [19]. Egg is an important source of several minerals, such as iron, phosphorous, zinc, copper, and other trace minerals. Regarding iron, one egg may supply 5.8% (USA) or 11.5% (Japan) of the recommended daily allowance (Table 4). The high calcium content of eggs is located almost completely in the shell, therefore, its direct utilization is rather difficult. However, modern shell processing technology has made shell calcium available as a food additive product. Recently, in a balance test using rats, we found that a proteinase-digested egg yolk protein increases the bioavailability of calcium in the diet.

III. FACTORS AFFECTING THE COMPOSITION OF HEN EGGS

Many variables directly affect the composition and concentration of nutrients in hen eggs. The concentration of lipids and cholesterol, as well as the degree of unsaturation of the lipids in eggs is significantly influenced by the strain of hen [21]. According to May and Stadelman, both water and protein content are also related to the strain [22]. Younger hens produce smaller eggs, and smaller eggs have a larger percentage of yolk, hence, age also affects egg composition [23]. Marion and his co-workers, as well as Menge and his co-workers found that lipid and cholesterol concentrations of yolk increased with age of the hens [24, 25]. In a study following the development of hens from 4 to 26 months, Cunningham and his co-workers found that age directly influenced the protein content and the concentration of phosphorous and chlorine of the laid eggs [19]. Egg nutrient composition is also directly affected by diet. It has been shown that dietary factors, such as the degree of unsaturation of the fat in hen's feed, modify

Table 5
Minerals in Hen Egg Compared with the Recomended Daily Allowance

	RDA (USA)[a]	RDA (Japan)[b]	Amount in an egg[c]	c/a×100 (%)
Calcium	1000 mg	600 mg	27 mg	2.7
Iron	18 mg	10 mg	1.15 mg	5.8
Magnesium	400 mg	300 mg	5.5 mg	1.4
Zinc	15 mg	15 mg	0.7 mg	4.7
Copper	2 mg	1.2-2.5 mg	0.08 mg	4
Manganese	-	3.4 mg	-	-
Sodium	nd	<10 g	66 mg	-
Potasium	1000 mg	<1.3 g	102.5 mg	10.3
Sulphur	nd	-	67 mg	-
Iodine	150 μg	6 μg	6 μg	4

[a] Recommended daily Allowances in the United States [3].

[b] Recommended daily Allowances for the Japanese (1994) [20].

[c] American egg board [17].

yolk cholesterol concentration and fatty acid composition [26, 27]. An increase of 22% in yolk cholesterol was found in eggs from hens on a diet supplemented with 30% linseed oil [28].

Jordan and his colleagues found that including corn oil in the diet of hens produced eggs with higher iodine value lipids [29]. Environmental temperature, and light and dark cycle conditions also have a direct impact on egg composition and, therefore, on the nutritional value. Hens exposed to higher environmental temperatures produced smaller eggs [30]. With proper handling, the nutritional quality of eggs is hardly affected. Egg processing technologies have little effect on the nutritional characteristics of the egg. According to the USDA Handbook 8-1 (USDA, 1976), frozen or liquid whole eggs are nutritionally identical to fresh whole eggs [1]. The same handbook suggests that drying might produce small losses of B vitamins. Drying had no effect on the digestibility of egg proteins. Dried eggs stored at 0°C for nine months showed no loss of vitamins B2 and D, niacin, or pantothenic acid [31].

Cooking improves digestibility and also influences nutrient composition. Heating denatures native proteins and enzyme inhibitors, helping to improve the availability of egg nutrients.

IV. SOME DRAWBACKS OF EATING EGGS

Most foods have some kind of drawbacks. The value of eggs as food is well recognized. However, the possibility of harmful substances in eggs has raised some concerns. The main adverse effects of eating eggs are that they might induce allergies, mainly in children. Another concern of hen eggs as food is their cholesterol content, which is detrimental for people with hypercholesterolemia.

A. ALLERGY INDUCED BY EGG PROTEIN

A certain minor portion of the protein of raw or slightly cooked eggs is absorbed directly into the blood stream, thus, it might be antigenic, especially to susceptible patients, frequently, infants. Allergy induced by hen eggs has been classified to be IgE-mediated and of immediate type (Type 1), which is known as atopy. The main allergens, so far reported, are ovalbumin, ovomucoid, and ovotransferrin in the egg white [32, 33]. Langeland, however, detected allergens even in egg yolk and identified them to be low-density lipoproteins and livetins [34]. Anet and his co-workers reported a correlation between the radio-allergosorbent test (RAST) results carried out for egg yolk and white, respectively [35]. A number of babies suffer from allergy to eggs, but, as time passes, most of them are gradually relieved of the trouble by the so-called 'oral tolerance' [36].

Heating decreases the allergenicity of egg proteins significantly [35], but its effectiveness depends to a great extent on the individual patient. This result indicates that the epitope of each allergic protein has its own sensitivity to heat. The allergenicity of ovomucoid is, however, known to be unaffected by heating [37].

B. BIOAVAILABILITY OF VITAMINS AND IRON

It is well known that avidin contained in egg white binds biotin to make this vitamin inactive. However, the possibility of a biotin deficiency is highly unlikely because this vitamin is found in various foods and, in addition, avidin is readily denatured by heat, becoming unable to bind the vitamin. In unheated egg albumen, Vitamin B12-binding protein limits the availability of the vitamin [38]. Although eggs also contain iron-binding proteins, the problem of the availability of iron has not been solved. For example, the question whether the iron bound to proteins, such as phosvitin and ovotransferrin, is absorbable or not, is still unanswered. The above two proteins have also been reported to decrease utilization of iron from other sources in the diet [39]. On the other hand, Sato and his co-workers [40] estimated the soluble and insoluble form of iron in the small intestine of rats fed a diet containing 20% of defatted egg yolk or soybean protein. They showed that, compared with soy protein, yolk protein increased the proportion of insoluble iron by about 20%. The contradictory results of several reports suggest that the absorption of iron in the presence of yolk is also affected by other factors.

C. CHOLESTEROL OF HEN EGG

Eggs are a source of high quality nutrients. However, lipid and cholesterol content has been labeled as a troublesome dietary factor. Egg cholesterol is found only in the yolk. For a long time, it was believed that the cholesterol content in one medium fresh egg was 274 mg. However, in 1988, the Egg Nutrition Center in the United States reported that the real cholesterol content was about 210 mg. The effect of dietary cholesterol on human health is still controversial. Although many animal experiments and numerous epidemiological and clinical observational surveys have been performed, it has not been possible to establish a direct relationship between dietary cholesterol and coronary heart disease [41, 42]. The amount of total cholesterol in circulating blood is only one of several risk factors that are associated with heart disease, and the composition of the diet is only one factor that affects the cholesterol levels in blood. Compared with dietary fat intake, cholesterol intake has a smaller effect in blood cholesterol. Nutritional and physiological trials have shown that dietary saturated fats contribute more than dietary cholesterol to increased serum cholesterol levels [43]. In spite of being the largest single source of cholesterol in the diet, normal consumption of eggs contributes only a small amount of saturated fat. For this reason, it is important to distinguish the effect of saturated fat on the diet from the effect of dietary cholesterol. Serum cholesterol levels are more critically affected by the intake of saturated and polyunsaturated fatty acids. According to Kritchevsky, the type and amount of protein and fiber influence blood cholesterol [44]. O'Dea and Sinclair reported that blood cholesterol levels of humans consuming two eggs daily were not elevated [45]. Farrell reported that consuming two eggs daily for 9 weeks produced no significant change in cholesterol or triglyceride levels in the plasma of volunteers [46]. Dawber et al. found no relationship between egg consumption and serum cholesterol and death from or incidence of heart disease [47]. The everyday diet already contains average amounts of dietary cholesterol and the increase of one or two eggs per day has no influence on serum cholesterol. Excluding eggs from the diet had no effect either [48]. In 1983, Liebman and Bazzarre pointed out that, "No relationship between egg cholesterol or total dietary cholesterol intakes and plasma lipid profiles could be discerned" and "care should be taken not to mislead the public into believing that simple restriction of eggs alone will result in significant decreases in plasma cholesterol levels" [49]. Although the cholesterol content of egg yolk is hardly altered by diet, the fatty acid content can be considerably modified through the hen's diet. Hens fed on diets high in unsaturated fats will produce egg yolks with more unsaturated fats. In recent years, efforts to enrich eggs with ω-3 fatty acids have been made. Jiang and Sim demonstrated that the cholesterolemic property of hen eggs could be reduced by incorporating ω-3 PUFA into the yolk [14]. The relative importance of ω-3 PUFA in the prevention of cardiovascular diseases is well known [50]. The cholesterol-lowering effect of ω-6 PUFA, such as linoleic acid, has been observed for eggs laid by hens fed on PUFA supplemented feed [51]. The manipulation of yolk lipid composition is well known. Therefore, it is possible to modify the fatty acid composition by adjusting the dietary lipid composition, making it viable to obtain eggs rich in linoleic acid, eicosapentaenoic acid (EPA), and docosahexaenoic acid (DHA) [52].

V. FINAL REMARKS

One egg contributes the same dietary requirements of protein as 35 g of meat. The nature and amount of total fat in the diet determine, to a large extent, the effect of eggs on blood cholesterol as do other factors [53]. Genetic character is the main factor in establishing the level of blood cholesterol and the way it is transported into the bloodstream. The life style of people (hypertension, smoking and drinking habits, diet, body weight, amount of exercise, etc.) is also directly related with the amount of cholesterol in the blood.

Many people, including egg consumers and scientists are confused and still relate dietary cholesterol directly to blood cholesterol levels. However, most recent research indicates that cholesterol from eggs has little effect on serum cholesterol.

In 1983, Harper observed that egg consumption in 1976, when heart disease rate was the minimum in the century, was around 16 kg per capita, very similar to 17 kg per capita in 1901-13, when the heart disease rate was almost at its maximum [54]. Decreasing dietary cholesterol by decreasing egg intake from five per week to two per week would be so small that this practice is not advisable [55]. A reduction in dietary cholesterol can de reached without elimination of eggs from the everyday meal.

In spite of the undeserved reputation of negative effects on human health, eggs will continue to be an inexpensive and complete source of essential nutrients. Furthermore, since the hen acts as a "biological filter", egg nutrients are normally free of contaminants and, therefore, safer for human consumption.

REFERENCES

1. **USDA,** Composition of foods. Dairy and egg products-raw, processed, prepared, in *Agric. Handbook*, USDA, Washington D. C., 1976.
2. **FAO/WHO/UNU.,** Energy and protein requirements, *Reports of the Joint FAO/WHO/UNU Expert Consultations (FAO, WHO and the United Nations University)*, Geneva, Switzerland, 1985
3. **National Research Council,** *Recommended Dietary Allowances*, Nat. Acad. Sci., Washington D. C., 1980.
4. **Yamaguchi, M. and Mastuno, N.,** Growth experiment in rats by use of purified whole egg protein, *Jpn. J. Nutr.*, **32**, 231, 1974.
5. **Zeisel, S. H.,** Choline: An important nutrient in brain development, liver function and carcinogenesis, *J. Am. Coll. Nutr.*, **11**, 473, 1992.
6. **Woodbury, M. M. and Woodbury., M. A.,** Neuropsychiatric development: Two case reports about the use of dietary fish oils and/or choline supplementation in children, *J. Am. Coll. Nutr.*, **12**, 239, 1993.
7. **Cossignani, L., Santinelli, F., Rosi, M., Simonetti, M. S., Valfre, F., and Damiani, P.,** Incorporation of ω-3 PUFA into hen egg yolk lipids. II: Structural analysis of triacylglycerols, phosphatidycholines and phosphatidyethanolamines, *Ital. J. Food Sci.*, **6**, 293, 1994.
8. **Jiang, Z., Ahn, D. U., and Sim, J. S.,** Effects of feeding flax and two types of sunflower seeds on fatty acid composition of yolk lipid classes, *Poult. Sci.*, **70**, 2467, 1991.
9. **Van Elswyk, M. E.,** Designer foods: Manipulating the fatty acid composition of meat and eggs for the health conscious consumer, *Nutr. Today*, **2**, 21, 1993.
10. **Simopoulos, A. P. and Salem, N.,** Egg yolk as a source of long-chain polyunsaturated fatty acids in infant feeding, *Am. J. Clin. Nutr.*, **55**, 411, 1992.
11. **Simopoulos, A. P.,** Fatty acids, in *Functional Foods-Designer Foods, Pharma Foods, Nutraceuticals*, Goldberg, I., Ed., Chapman & Hall, New York, 1994, p 355.
12. **Neuringer, M. E. and Connor, W. E.,** ω-3 fatty acids in the brain and retina: Evidence for their essentiality, *Nutr. Rev.*, **44**, 285, 1986.
13. **Simopoulos, A. P.,** ω-3 fatty acids in health and disease and in growth and development, *Am. J. Clin. Nutr.*, **54**, 438, 1991.
14. **Jiang, Z. and Sim, J. S.,** Effect of dietary ω-3 PUFA-enriched chicken eggs on plasma and tissue cholesterol and fatty composition of rats, *Lipids*, **27**, 279, 1992.
15. **Kokestu, M., Nitoda, T., Juneja, L. R., Kim, M., Kashimura, N., and Yamamoto, T.,** Sialyloligosaccharides from egg yolk as an inhibitor of rotaviral infection, *J. Agric. Food Chem.*, **43**, 858, 1995.
16. **Koketsu, M., Nakata, K., Juneja, L. R., Kim, M., and Yamamoto, T.,** Learning performance of egg yolk sialyloligosaccharide fraction, *Oyo Toshitu Kagaku (in Japanese)*, **42**, 15, 1995.
17. **American Egg Board.,** A scientist speaks about eggs, *American Egg Board*, 1461 Renaissance Drive, Suite 301, Park ridge, IL 60068, 1974.
18. **Jiang, Y. H., McGeachin, R. B., and Bailey, C. A.,** α-tocophenol, β-carotene, and retinol enrichment of chicken eggs, *Poult. Sci.*, **73**, 1137, 1994.
19. **Cunningham, F. E., Cotterill, O. J., and Funk, E. M.,** The effect of season and age of bird. 2: On the chemical composition of egg white, *Poult. Sci.*, **39**, 300, 1960.
20. **Japanese Association of Nutriologist.,** Nutritional physiology, in *Indispensable Manual for Nutriologist*, Daiichi Shuppan Co., Ltd., Tokyo, 1994.
21. **Edwards, H. M., Jr., Driggers, J. C., Dear, R., and Carmon, J. L.,** Studies on the

cholesterol content of eggs from various breeds and/or strains of chickens, *Poult. Sci.*, **39**, 487, 1960.
22. **May, K. N. and Stadelman, W. J.**, Some factors affecting components of eggs from adult hens, *Poult. Sci.*, **39**, 560, 1960.
23. **Forsythe, R. H.**, Chemical and physical properties of eggs and egg products, *Cereal Sci. Today*, **8**, 309, 1963.
24. **Marion, J. E., Woodroff, J. G., and Tindell, D.**, Physical and chemical properties of eggs as affected by breeding and age of hens, *Poult. Sci.*, **45**, 1189, 1966.
25. **Menge, H., Littlefield, L. H., Forbish, L. T., and Weinland, B. T.**, Effect of cellulose and cholesterol on blood and yolk lipids and reproductive efficiency in the hen, *J. Nutr.*, **104**, 1554, 1974.
26. **Hargis, P. S.**, Modifying egg yolk cholesterol in domestic fowl, a review, *World Poult. Sci. J*, **44**, 17, 1988.
27. **Stadelman, W. J. and Pratt, D. E.**, Factors influencing composition of the hen's egg, *World Poult. Sci. J.*, **45**, 247, 1989.
28. **Weiss, J. F., Naber, E. C., and Johnson, R. M.**, Effect of a dietary fat and other factors on egg yolk cholesterol. The 'cholesterol' content of egg yolk as influenced by dietary unsaturated fat and the method of determination, *Arch. Biochem. Biophys.*, **105**, 521, 1964.
29. **Jordan, R., Vail, G. E., Rogler, J. C., and Stadelman, W. J.**, Further studies on eggs from hens on diets differing in fat content, *Food Technol.*, **16**, 118, 1962.
30. **Carmon, L. G., and Houston, T. M.**, The influence of environmental temperature upon egg components of domestic fowl, *Poult.Sci.*, **44**, 1237, 1965.
31. **Kramer, A.**, Effect of storage on nutritive value of food, in *Handbook of Nutritive Value of Processed Food*, Rechcigl, M. J., Ed., CRC Press, Inc., Boca Raton, FL., 1982.
32. **Langeland, T.**, A clinical and immunological study of allergy to hen's egg white. II. Antigens in hen's egg white studies by crossed immunoelectrophoresis, *Allergy*, **37**, 323, 1982.
33. **Hoffman, D. R.**, Immunochemical identification of the allergens in egg white, *J. Allergy Clin. Immunol.*, **71**, 481, 1983.
34. **Langeland, T.**, A clinical and immunoligial study of allergy to hen's egg white. IV. Specific IgE antibodies to individual allergens, *Allergy*, **38**, 493, 1983.
35. **Anet, J., Back, J. F., Baker, R. S., Barnett, D., Burley, R. W., and Howden, M. E. H.**, Allergens in the white and yolk of hen's egg, *Int. Arch. Allergy Appl. Immunol*, **77**, 364, 1985.
36. **Richman, L. K., Graeff, A. S., Yarchoan, R., and Strober, W.**, Simultaneous induction of antigen-specific IgA helper T cells and IgG suppressor T cells in the Peyer's patch after protein feeding, *J. Immunol*, **126**, 2079, 1981.
37. **Konishi, Y., Kurisaki, J., Kaminogawa, S., and Yamaguchi, K.**, Determination of antigenicity by radioimmunoassay and of trypsin inhibitory activities in heat or enzyme denatured ovomucoid, *J. Food Sci.*, **50**, 1422, 1985.
38. **Levine, A. S. and Doscherholmen, A.**, Vitamin B12 bioavailability from egg yolk and egg white; relationship to binding proteins, *Am. J. Clin. Nutr.*, **38**, 436, 1983.
39. **Callender, S. T.**, Egg and iron absorption, *Brit. J. Haematol.*, **19**, 657, 1970.
40. **Sato, R., Lee, Y. S., Noguchi, T., and Naito, H.**, Iron solubility in the small intestine of rats fed egg yolk protein, *Nutr. Reports Int.*, **30**, 1319, 1984.
41. **Texon, M.**, The cholesterol-heart disease hypothesis (critique) time to change course?, *Bull. N. Y. Acad. Med.*, **65**, 836, 1989.
42. **Stehbens, W. E.**, The controversial role of dietary cholesterol and hypercholesterolemia in coronary heart disease and atherogenesis, *Pathology*, **21**, 213, 1989.

43. **Moore, T. J.,** The cholesterol myth, *Atlantic*, September 1989, 37, 1989.
44. **Kritchevsky, D.,** Dietary interactions, in *Nutrition, Lipids and Coronary Heart Disease*, Levy, R. E., Dennis, B. H., Rifkind, B. M., and Ernst, N., Eds., Raven Press, New York, 1979, p 229.
45. **O'Dea, K. and Sinclair, A. J.,** Eggs, implications for human health, in *Australian Poultry Science Symposium, University of Sydney, NSW*, 1991, p 23.
46. **Farrell, D. J.,** The fortification of hen's eggs with ω-3 long chain fatty acids and their effect in humans, in *Egg Uses and Processing Technologies. New developments*, Sim, J. S. and Nakai, S., Eds., CAB International, Wallingford, UK, 1994, p 386.
47. **Dawber, T. R., Nickerson, R. J., Brand, F. N., and Pool, J.,** Eggs, serum cholesterol and coronary heart disease, *Am. J. Clin. Nutr.*, **36**, 616, 1982.
48. **Flynn, M. A., Nolph, G. P., Flynn, T. C., Kahrs, R., and Krause, G.,** Effect of dietary egg on human serum cholesterol and triglyceride, *Am. J. Clin. Nutr.*, **32**, 1051, 1979.
49. **Liebman, M. and Bazarre, T.,** Plasma lipids of vegetarian and nonvegetarian males: Effects of egg consumption, *Am. J. Clin. Nutr.*, **38**, 612, 1983.
50. **Dryberg, J., and Jorgensen, K. A.,** Marine oils and thrombogenesis, *Prog. Lipid Res. Q*, **46**, 40, 1982.
51. **Grundy, S. M.,** Comparison of monounsaturated fatty acids and carbohydrates for lowering plasma cholesterol, *N. Engl. J. Med.*, **314**, 745, 1986.
52. **Yu, M. M. and Sim, J. S.,** Biological incorporation of ω-3 fatty acids into chicken egg, *Poult. Sci.*, **66**, 195, 1987.
53. **Reiser, R.,** Egg cholesterol in human health, in *Proceedings of the World's Poultry Science Congress, Nagoya, Japan*, 1988, 16.
54. **Harper, A. E.,** Coronary heart disease - An epidemic related to diet?, *Am. J. Clin. Nutr.*, **37**, 669, 1983.
55. **Trunswell, A. S.,** Diet and plasma lipids - A reappraisal, *Am. J. Clin. Nutr.*, **21**, 977, 1978.

Chapter 4

INSIGHTS INTO THE STRUCTURE-FUNCTION RELATIONSHIPS OF OVALBUMIN, OVOTRANSFERRIN, AND LYSOZYME

H R. Ibrahim

TABLE OF CONTENTS

I. Introduction
II. Structure and Function of Ovalbumin
    A. Structural Properties of Ovalbumin
    B. Structural Resemblance Between Ovalbumin and Proteinase Inhibitors
    C. Other Structural Functional Aspects of Ovalbumin
III. Structure and Function of Ovotransferrin
    A. Structural Properties of Ovotransferrin
    B. Iron Transport by Ovotransferrin
    C. Antimicrobial Activity of Ovotransferrin
IV. Structure and Function of Lysozyme
    A. Structural Properties of Lysozyme
    B. Biological Functions of Lysozyme
    C. Conformational Transition of Lysozyme
    D. Thermal Stability and Denaturation
V. Concluding Remarks

References

I. INTRODUCTION

The protein content of an egg is 12-13% by weight, mainly distributed through three parts: egg membranes, egg white, and egg yolk. These parts consist of approximately 4, 12, and 31 % by weight of protein, respectively. Most of egg white proteins are soluble and can easily be isolated. Vadehra and Nath have reported that the egg white contains approximately 40 different proteins [1]. It has long been believed that these distinct sorts of proteins play important roles during embryogenesis and deterioration of the egg. The physicochemical properties of egg white proteins with vital functional properties might be expected to undergo structural changes during the embryogenesis process. These macromolecules exist as basic subunits or structures which can complex homogeneously or heterogeneously with each other to form more intricate macromolecular systems or morphological entities. Although no definite biological function has yet been found for egg white proteins, they possess unique functional properties, such as antimicrobial, enzymatic and antienzymatic, cell growth stimulatory, metal binding, vitamin binding, and immunological activities. A structural feature of egg white proteins is that most of them are glycoproteins in which the carbohydrate moieties are either glycosidically linked to serine or threonine residues (O-linked sugar), or attached to an asparagine residue through an amide linkage (N-linked sugar). They also exist in different, distinctive structural entities, e.g. phosphoglycoproteins and sulfated glycoproteins.

    Egg white constitutes the second line of defense against invading bacteria next to the shell and shell membranes of the eggs. Egg white proteins have been studied by a vast spectrum of scientists from different disciplines using them as the source of typical (globular) proteins. Egg white proteins have different dynamic structures and their amphiphilic nature possesses various physiological and biotechnological functions. The biological functions of egg white proteins have been thought to prevent the penetration of microorganisms into the yolk and to provide nutrients to the embryo during the late stages of development. Of the many different types of proteins found in egg albumen, most of them appear to possess antimicrobial properties or certain physiological functions to interfere with the growth and spread of invading microorganisms. As shown in Table 1, these functions are (i) proteinase inhibitors, such as ovoinhibitor, ovomucoid, ovomacroglobulin, and cystatine; (ii) bacteriolytic enzymes, e.g. lysozyme; (iii) vitamin chelating proteins, e.g., avidin and ovoflavoprotein; (iv) metal chelator proteins, e.g., ovotransferrin; and (v) jelly-like protein, e.g., ovomucin.

Recent articles have focused on evidence bearing on the biological functions of proteins from different sources. In this sense, determination of the structure-function relationships will allow many observations on the chemistry of egg white proteins to be placed in a clearer structural functional perspective. This chapter summarizes the current information on three structurally and functionally different globular proteins of egg white proteins to help establishing structure-function relations and to clarify possible roles for their existence in the hen egg. The proteins discussed in this chapter have been selected because they are good candidates for elucidating the structure-function correlation of a protein. One of them is an intriguing protein with unknown biological function (ovalbumin), another is an iron-binding protein with unclear antimicrobial properties (ovotransferrin), and the third is a bacteriolytic protein with well-characterized physicochemical properties, though we still have many approaches to be accomplished to improve our understanding of its biological function (lysozyme). These are the subjects of this review [2-26].

II. STRUCTURE AND FUNCTION OF OVALBUMIN

A. STRUCTURAL PROPERTIES OF OVALBUMIN

Ovalbumin, a monomeric glycoprotein with a molecular weight of 44,000 Da and isoelectric point of 4.7, constitutes more than 50 % of the total egg white proteins. It has been the subject of physical chemical studies for few decades as a convenient model of protein. Despite the intensive investigations undertaken on ovalbumin, its function is unknown. The chemical and physical properties have recently been reviewed [27]. The amino acid sequence of hen egg white ovalbumin, comprising 385 residues, was deduced from the cloned DNA [28] and by peptide sequencing of purified protein [29]. Nearly 50 % of the residues of ovalbumin are hydrophobic [27]. The sequence includes four sulfhydryl groups with a single disulfide bridge which has once been reported to exist between Cys 73 and 382 [30], but more recent research found that the single disulfide bond exists between Cys 73 and 120 [29]. In fact, these conflicting results argue that sulfhydryl-disulfide interchange reaction among the free thiol groups may trigger a certain functional conformation of ovalbumin which has not been discovered, so far. Particularly, when an egg is stored in the cold, a proportion of the ovalbumin progressively transforms into a more heat stable form referred to as S-ovalbumin [31]. Although this phenomenon is of great interest to biochemists and biologists, it has been poorly characterized. From this viewpoint, it has been proposed that disulfide interchanges are involved in the conversion of ovalbumin to S-ovalbumin on storage [31]. Four thiol groups of ovalbumin are chemically reactive only when the molecule is denatured, most probably, because an ovalbumin molecule is compactly folded, but not highly cross-linked. Ovalbumin has an acetylated *N*-terminus glycine, a *C*-terminus proline, and a single heterogeneous carbohydrate chain covalently linked to the amide nitrogen of Asn 292. Different glycopeptides have also been

Table 1
Physiological and Functional Properties of Egg White Proteins

Protein	Function
Ovalbumin	Lacks the transient *N*-terminal signal sequence "leader peptide" [2]. Induces fusion of phospholipid vesicles at low pH. [3,4] Useful model for studying structure-function relationship of proteins. Its proteolytic peptides usually fold into one molecule essentially the same as the native conformation [5, 6]. A suitable model for studying factors affecting the rate of protein synthesis *in vivo* using oviduct tissue.
Ovotransferrin	Homologous to serum transferrin and lactoferrin [7]. Complexes Fe^{3+}, Zn^{2+}, Al^{3+}, Cu^{2+} with high affinity. Antimicrobial protein, particularly, against gram negative bacteria. It has specific receptor on the surface of chick-embryo red cell [8].
Ovomucoid	Specific trypsin inhibitor 1:1 complex. Inhibits bacterial and fungal proteinases. Antimicrobial protein [9,10,11,12].
Ovoinhibitor	Proteinase inhibitor with wide range of specificities; inhibits trypsin, chymotrypsin, elastase, subtilisin, and alkaline proteinases [13,14].
Lysozyme	Bacteriolytic enzyme lyses the glycans in the cell walls of gram positive bacteria [15]. Strongly interacts with the outer membrane lipopolysaccharides of gram negative bacteria [16]. Strongly interacts with ovomucin, thus, increasing the viscosity of albumen [17]. Weak nonspecific esterase activity [15]. Weak proteolytic activity [18]. Antiviral, antitumor activities [19].
Ficin inhibitor (Cystatin)	Inhibitor of cysteine proteinases, e.g., ficin, papain, cathepsin C1, cathepsin B1 [20]. Bacteriostatic to microorganisms producing cysteine proteinases.
Avidin	Strongly interacts with biotin [21]. Deprives the essential biotin for bacterial growth (bacteriostatic). Acts as a biotin carrier for the chicken embryo.
Ovoflavoprotein	Moderately binds riboflavin (vitamin B2) [22]. Act as a carrier and store of riboflavin for chick embryo.
Ovomucin	Determines the viscosity of albumen (mucilagenous nature), thus, serves to prevent the spread of microorganisms. Antiviral activities. Associates with different proteins in the albumen, e.g., lysozyme and β-*N*-acetylglucosaminidase. Both proteins have particular antimicrobial activity.
Ovomacroglobulin	Inhibits a wide range of proteinases [23,24].
Glutamyl aminopeptidase	Highly specific proteinase for α-glutamyl or α-aspartyl residues [25].
β-*N*-acetylglucos-aminidase	Inhibits the growth of gram negative bacteria [26].

identified in ovalbumin. All share a common core structure: Mannose β(1-4)GlcNAc-β(1-4)GlcNAc-Asn298 [32]. Three main ovalbumin forms exist with two, one, and zero phosphate groups per molecule at Ser 68 and Ser 344 [29], and the dephosphorylated form constitutes 80 % of the total ovalbumin. The sequence 346-352, comprising the phosphoserine 344, occupies a particularly exposed region on the surface of the molecule [33].

B. STRUCTURAL RESEMBLANCE BETWEEN OVALBUMIN AND PROTEINASE INHIBITORS

Treatment of ovalbumin with subtilisin under defined conditions releases a hexapeptide (Ala-Gly-Val-Asp-Ala-Ala) at the residue 358, leaving two peptide chains strongly bound by noncovalent bonds. This form is called plakalbumin [34]. Ovalbumin and plakalbumin show structural homology with a family of serine proteinase inhibitors called the serpins superfamily. The serpins family, a superfamily of more than 20 homologous proteins found in animals, plants, and viruses, that probably evolved from a serine proteinase inhibitor, includes human antitrypsin, α–1-proteinase inhibitor (A1PI), antithrombin III, and antiplasmin [35]. The stereoview drawings of plakalbumin and A1PI are compared in Figure 1, showing the structural homology between the two proteins. The secondary structural elements of the two proteins are also compared, manifesting remarkable similarities in the β-sheets, and conservation in the turns, bulges, and α–helix. More detailed information on the structure of plakalbumin and A1PI is emphasized in the literature [6], and the X-ray crystallographic analysis of intact ovalbumin can also be found elsewhere [33]. Ovalbumin shows no inhibitory activity, but some members of serpins family have been shown to be sensitive to proteolytic cleavage at residue 358 in the sequence where subtilisin cleaves ovalbumin to form plakalbumin [36]. It should be noted that the released segment from ovalbumin by subtilisin is in a region very close to phosphoserine 344. Thus, the phosphorylation-dephosphorylation may also be critical for the conversion to the biologically active form. The reported alkaline phosphatase activities in the hen's egg vitelline membrane agree well with this assumption [37]. Future studies on the possible formation of plakalbumin *in vivo* and the discovery of new enzymatic activities in the minor proteins of egg white may provide a clue to the biological function of ovalbumin.

C. OTHER STRUCTURAL FUNCTIONAL ASPECTS OF OVALBUMIN

Interesting research has recently revealed that reduced ovalbumin is hydrolyzed by subtilisin into three major fragments, and the three fragments fold into one molecule showing the same CD spectra and retention time from gel permeation chromatography as native ovalbumin [5]. An additional approach to the function of ovalbumin might come from the finding that ovalbumin induces the fusion of acidic phospholipid vesicles, particularly at low pH [4]. In the egg, ovalbumin was found to be absorbed in its intact form by the yolk sac during the process of embryogenesis, although it is unknown whether its absorption is specific or not. The absorption process would require fusion and translocation competency.

The way in which ovalbumin denatures is of particular interest. Changes in the secondary structural elements upon heat denaturation of ovalbumin was studied by using circular dichroism analysis [38]. The critical effect of intermolecular hydrophobic interactions on the promoted β-sheet contents was emphasized. Change in the CD spectra during and after heating was interpreted as the result of an increase in β-sheet structure with a concomitant decrease of the helix content of the molecule. The secondary structural changes of an ovalbumin molecule after heating at different temperatures followed by immediate cooling to 25°C are shown in Figure 2. In this essence, ovalbumin is known to form a transparent standing gel upon heating for gelation, the characteristics of the organized thread of beads' macromolecular interactions.

Although many papers have been published on ovalbumin during the past few years, its biological role in the egg remains unknown. It is believed that ovalbumin, especially its unphosphorylated form, serves as a source of amino acids for the embryo. What about the phosphorylated forms?

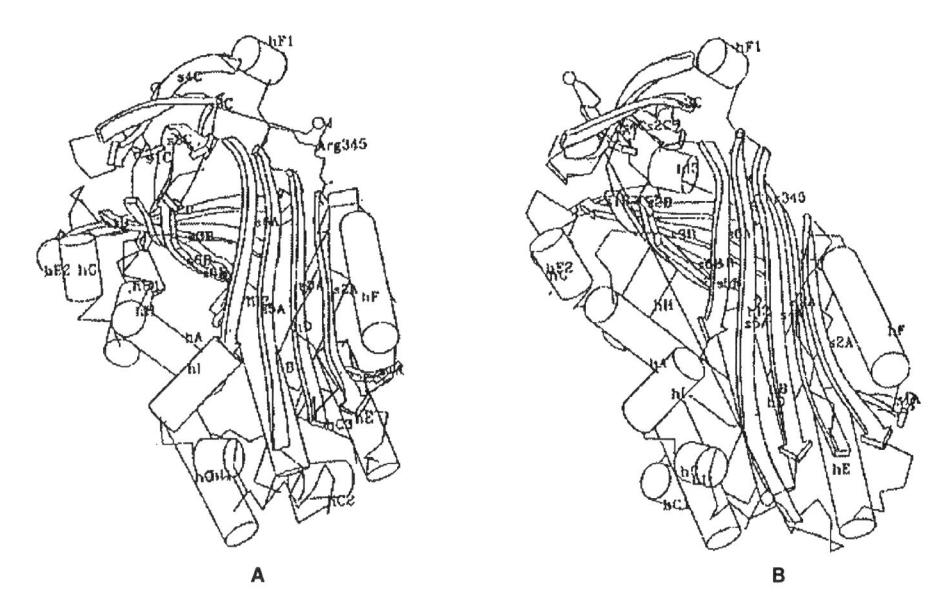

Figure 1. Stereo drawing of the secondary structural elements of (A) plakalbumin and (B) A1PI. Cylinders represent helices (h), and arrows represents β-strands(s). The cleaved segments are indicated by long, straight connections and open circles. The Arg and Thr 3 45 side chains are also shown. Reproduced from Wright et al. (1990) [6] by permission of Academic Press, Ltd., London.

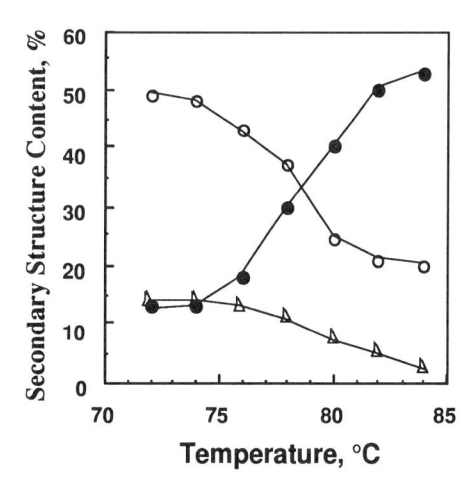

Figure 2. Changes in the content of the secondary structure elements of ovalbumin during heat denaturation. A 0.1% ovalbumin solution was made at the indicated temperature and then immediately cooled to 25°C. o-o , α-helix; ●-● , β-sheet; ∆-∆ , β-turn structure. Reproduced from Kato and Takagi (1988) [38] by permission of American Chemical Society, Washington, D. C.

(A)

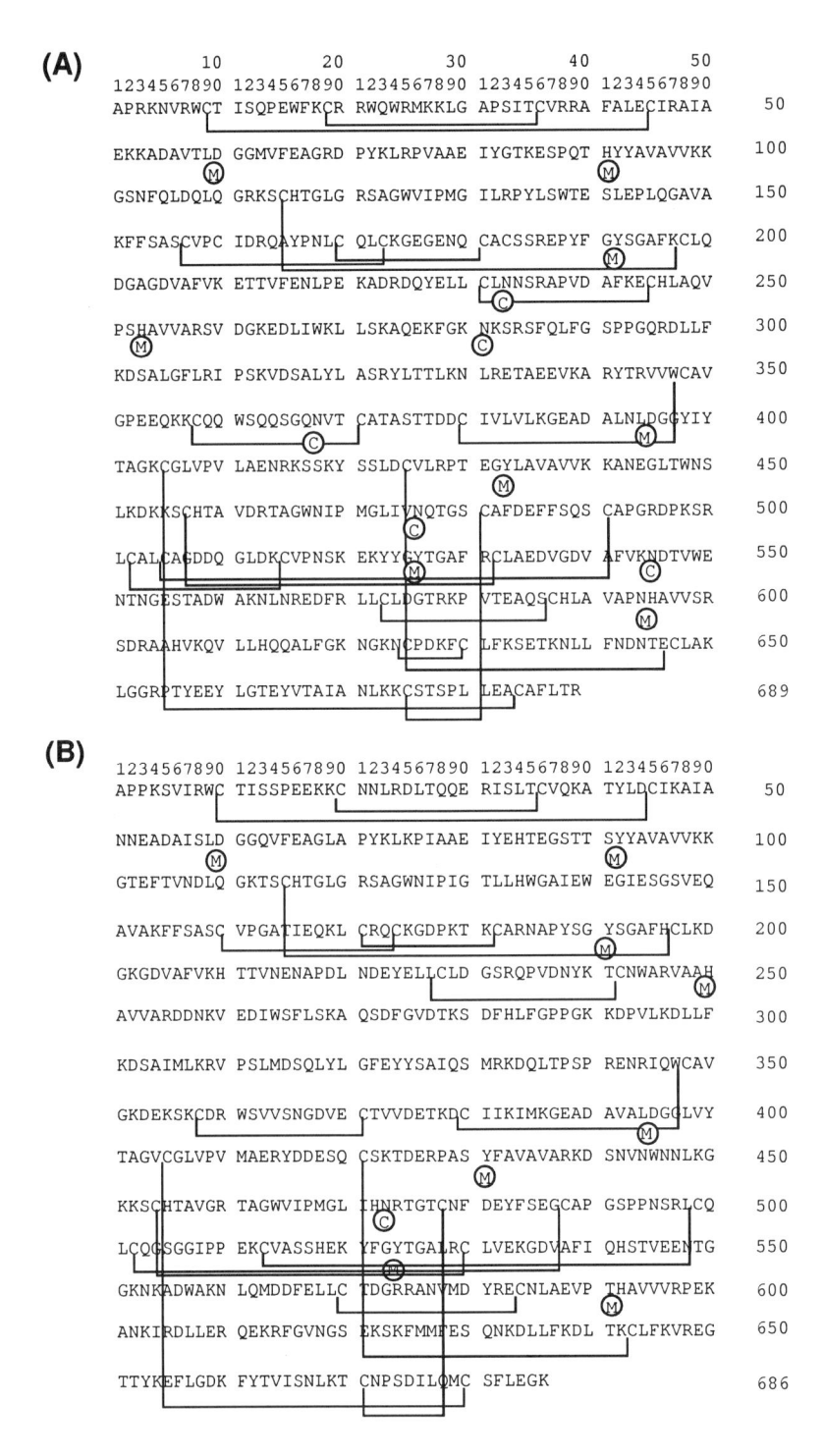

```
             10          20          30          40          50
    1234567890 1234567890 1234567890 1234567890 1234567890
    APRKNVRWCT ISQPEWFKCR RWQWRMKKLG APSITCVRRA FALECIRAIA    50
    EKKADAVTLD GGMVFEAGRD PYKLRPVAAE IYGTKESPQT HYYAVAVVKK   100
    GSNFQLDQLQ GRKSCHTGLG RSAGWVIPMG ILRPYLSWTE SLEPLQGAVA   150
    KFFSASCVPC IDRQAYPNLC QLCKGEGENQ CACSSREPYF GYSGAFKCLQ   200
    DGAGDVAFVK ETTVFENLPE KADRDQYELL CLNNSRAPVD AFKECHLAQV   250
    PSHAVVARSV DGKEDLIWKL LSKAQEKFGK NKSRSFQLFG SPPGQRDLLF   300
    KDSALGFLRI PSKVDSALYL ASRYLTTLKN LRETAEEVKA RYTRVVWCAV   350
    GPEEQKKCQQ WSQQSGQNVT CATASTTDDC IVLVLKGEAD ALNLDGGYIY   400
    TAGKCGLVPV LAENRKSSKY SSLDCVLRPT EGYLAVAVVK KANEGLTWNS   450
    LKDKKSCHTA VDRTAGWNIP MGLIVNQTGS CAFDEFFSQS CAPGRDPKSR   500
    LCALCAGDDQ GLDKCVPNSK EKYYGYTGAF RCLAEDVGDV AFVKNDTVWE   550
    NTNGESTADW AKNLNREDFR LLCLDGTRKP VTEAQSCHLA VAPNHAVVSR   600
    SDRAAHVKQV LLHQQALFGK NGKNCPDKFC LFKSETKNLL FNDNTECLAK   650
    LGGRPTYEEY LGTEYVTAIA NLKKCSTSPL LEACAFLTR              689
```

(B)

```
    1234567890 1234567890 1234567890 1234567890 1234567890
    APPKSVIRWC TISSPEEKKC NNLRDLTQQE RISLTCVQKA TYLDCIKAIA    50
    NNEADAISLD GGQVFEAGLA PYKLKPIAAE IYEHTEGSTT SYYAVAVVKK   100
    GTEFTVNDLQ GKTSCHTGLG RSAGWNIPIG TLLHWGAIEW EGIESGSVEQ   150
    AVAKFFSASC VPGATIEQKL CROCKGDPKT KCARNAPYSG YSGAFHCLKD   200
    GKGDVAFVKH TTVNENAPDL NDEYELLCLD GSRQPVDNYK TCNWARVAAH   250
    AVVARDDNKV EDIWSFLSKA QSDFGVDTKS DFHLFGPPGK KDPVLKDLLF   300
    KDSAIMLKRV PSLMDSQLYL GFEYYSAIQS MRKDQLTPSP RENRIQWCAV   350
    GKDEKSKCDR WSVVSNGDVE CTVVDETKDC IIKIMKGEAD AVALDGGLVY   400
    TAGVCGLVPV MAERYDDESQ CSKTDERPAS YFAVAVARKD SNVNWNNLKG   450
    KKSCHTAVGR TAGWVIPMGL IHNRTGTCNF DEYFSEGCAP GSPPNSRLCQ   500
    LCQGSGGIPP EKCVASSHEK YFGYTGALRC LVEKGDVAFI QHSTVEENTG   550
    GKNKADWAKN LQMDDFELLC TDGRRANVMD YRECNLAEVP THAVVVRPEK   600
    ANKIRDLLER QEKRFGVNGS EKSKFMMFES QNKDLLFKDL TKCLFKVREG   650
    TTYKEFLGDK FYTVISNLKT CNPSDILQMC SFLEGK                 686
```

Figure 3. Primary structures of bovine lactoferrin (A) and ovotransferrin (B). Disulfide cross-links (line), sites for sugar attachment (C), and residues involved in the iron-binding sites (M) are shown. Figure 3 (A) and (B), reproduced from Pierce et al. (1991) [42] and Williams et al. (1982) [43], respectively, by permission of Springer-Verlag Heidelberg, Heidelberg.

III. STRUCTURE AND FUNCTION OF OVOTRANSFERRIN

A. STRUCTURAL PROPERTIES OF OVOTRANSFERRIN

Ovotransferrin (conalbumin) is a monomeric glycoprotein of molecular weight 78,000, contains no free sulfhydryl groups or phosphorus, and has the capacity to bind reversibly with two Fe^{3+} ions per molecule and concomitantly with two CO_2^- ions. It is a member of a group of three closely related iron-binding proteins. Three proteins typify the transferrin family: serum transferrin (iron transport protein), lactoferrin (in milk and other secretions), and ovotransferrin (in egg white) [39, 40]. The amino acid sequences of ovotransferrin and lactoferrin are compared in Figure 3, showing the position of disulfide bonds, the amino acid residues involved in the iron-binding sites (M), and the sites of the carbohydrate moieties of both proteins. Although the two proteins show remarkable similarity in amino acid sequence and the global folding of the molecules, they differ in their disulfide cross-linking patterns. Moreover, the number of attached carbohydrate chains is also different. The iron-binding sites are in the same position in both proteins, where they comprise the two phenolate oxygens of two tyrosines and a carboxylate oxygen from aspartic acid, and one neutral nitrogen, the imidazole nitrogen of a single histidine. Analysis of the structure of transferrins shows that the single polypeptide chain of 680-700 residues can be divided into two homologous halves, each with a single iron site (a bilobal molecule) [41-44]. The two lobes are connected by an α–helix segment 10 amino acids long. Each lobe is further divided into two domains of about 160 residues each with an iron site between them. The most notable differences between the transferrin members originate from the number and the location of the glycosylation sites.

B. IRON TRANSPORT BY OVOTRANSFERRIN

The significant structural similarities between lactoferrin and ovotransferrin justify the similarity of their biological roles, implication in the transport of iron in a soluble form to the target cells. The recognition of transferrin molecules by the target cells is mediated by membrane-bound receptors (transferrin receptor). In animal tissues, lactoferrin binds specific receptors on platelet cells, thus, inducing various immunological responses. For ovotransferrin, its binding to chick-embryo red cell has been reported [8]. A recent study showed that, after cleavage of the connecting peptide (between the *N*-terminal, Nt, and C-terminal, Ct, lobes of the molecule) of ovotransferrin with trypsin, the two lobes, 30 kDa Nt and 50 kDa Ct, remain attached to each other (noncovalently), showing cooperative interactions [45]. In parallel, a related investigation showed that Nt or Ct half-molecule fragments alone cannot bind the chick red blood cells, but, when both fragments are present, the binding and delivery of iron is observed [46]. The conclusions from these studies were that the reassociation of the half-molecule fragments has a significant function which allows the complex to be recognized by the transferrin receptor. Moreover, it was found that an iron-saturated Nt or NC fragment will not bind tightly to the receptor until a contralateral fragment is present, and it does not matter whether the latter is carrying Fe^{3+} or not.

C. ANTIMICROBIAL ACTIVITY OF OVOTRANSFERRIN

The high affinity of transferrins for iron means that, in the presence of unsaturated transferrin, (apotransferrin) iron will be sequestered and rendered unavailable for the growth of microorganisms. The antimicrobial activity of ovotransferrin [47, 48] and lactoferrin against *E. coli* and many other bacteria has been reported [49]. The antimicrobial action was attributed to the iron sequestration demonstrated against a variety of microorganisms, including pathogenic *E. coli*, *Pseudomonas aeruginosa*, and *Vibrio cholera* [49]. Although the studies so far reported attributed the antimicrobial effects of lactoferrin and ovotransferrin to be bacteriostatic (growth inhibiting), the bactericidal action of lactoferrin has recently been reported against a wide spectrum of microorganisms [50]. This bactericidal action was related to the bactericidal sequence of 25 amino acid residues in the N-terminal region between residues 37 and 61 of the lactoferrin molecule. This peptide fragment was isolated by hydrolysis with pepsin and has been named lactoferricin. The purified lactoferricin exhibited greater bactericidal activity against a wide spectrum of microorganisms than the intact lactoferrin [50]. The primary structures of lactoferricin and its corresponding sequence from ovotransferrin are shown in Figure 4.

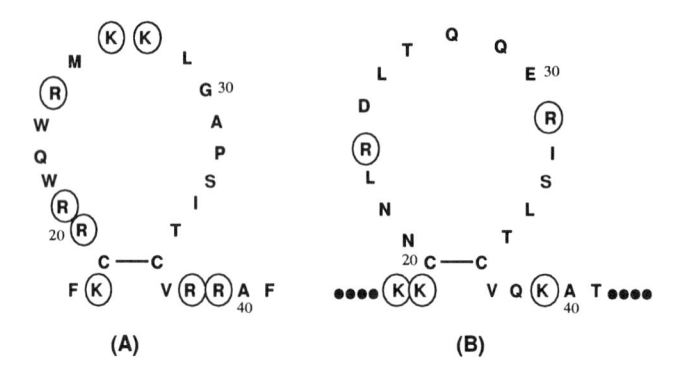

Figure 4. Primary structures of bovine lactoferricin (A) and its corresponding sequence in ovotransferrin (B). Positively charged residues are circled. Figure 4 (A), reproduced from Bellamy et al. (1992) [50] by permission of Elsevier Science Publishers B. V., Amsterdam.

The loop structure, imposed by a disulfide bond between Cys 19 and Cys 36 of lactoferrin, and the high basicity of the lactoferricin contributes to the observed bactericidal activity and to the direct membrane damage of the exposed bacteria. For ovotransferrin, the loop structure exists but the net positive charge of the segment is less than that of lactoferricin. Isolation of the corresponding sequence from ovotransferrin was attempted in our laboratories to compare its antimicrobial properties with that reported for lactoferricin. We found that the peptide fragment can not be obtained by pepsin treatment. Sequential homological analysis revealed inconsistency between ovotransferrin and lactoferrin in the amino acid sequence in the region harboring the lactoferricin peptide. Therefore, the peptic peptides from ovotransferrin were largely different from those obtained from lactoferrin. In addition, the peptic hydrolysate of ovotransferrin showed no antimicrobial activity, but the hydrolysate of lactoferrin was strongly bactericidal. Selection of the proteinase for this investigation would be of great importance, and screening of different proteinases are now in progress. However, the remarkable structural similarity between lactoferrin and ovotransferrin, with the exception of their different net positive charges and their carbohydrate contents, argue that ovotransferrin possesses structurally dependent bactericidal activity other than the iron deprivation effect like that reported for lactoferrin. To verify this assumption, we have examined the antimicrobial activity of the diferric- (OTf-Fe2), monoferric- (OTf-Fe1), and iron-free (apo-OTf) ovotransferrin against *E. coli* K12 and *Staphylococcus aureus* (Figure 5). Regardless of the degree of iron saturation, ovotransferrin exhibited strong antimicrobial action against gram positive *St. aureus*. For gram negative *E. coli*, a very interesting observation was obtained that the iron-bound ovotransferrin was more potent bactericidally than the iron-free ovotransferrin (apo-OTf). It is clear, therefore, that ovotransferrin is bactericidal and the mechanism of action can be attributed to structural factors rather than the iron deprivation effect. Presumably, a certain conformation, which occurs when iron is bound to the molecule of ovotransferrin, is responsible for the antimicrobial activity against gram negative bacteria. This idea agrees well with a recent report showing bactericidal synergism between lysozyme and ovotransferrin in the egg white. The phenomenon is of great interest and may explain the function of ovotransferrin in the avian egg. More detailed information on the structural factors accounting for the bactericidal activity of ovotransferrin may uncover interesting structure-biological function relationships for this protein. Comparison between the structure and the mode of bactericidal action of lactoferrin and ovotransferrin is currently in progress in our laboratories, to provide a clue for elucidating the yet unknown functions of ovotransferrin. Understanding the role of this protein and the other antimicrobial proteins, such as lysozyme, proteinase inhibitors, and aminopeptidase, may help in clarifying the molecular mechanism of the antimicrobial complex of the hen egg white.

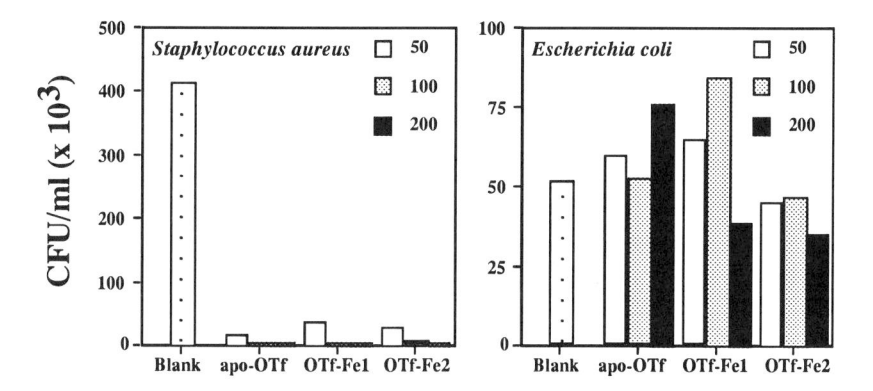

Figure 5. Antimicrobial effects of iron-free ovotransferrin (apo-OTf), monoferric ovotransferrin (OTf-Fe1), and diferric ovotransferrin (OTf-Fe2) against *Staphylococus aureus* and *Escherichia coli.* Assay was performed at 50, 100, and 200 μg protein per ml of mid-log phase cell suspension (OD 675 = 0.0005) made in 1% Bacto-peptone. Incubation of protein with bacteria lasted for 1hr at 37°C before determining the colony forming unit (CFU) on nutrient agar plates.

IV. STRUCTURE AND FUNCTION OF LYSOZYME

A. STRUCTURAL PROPERTIES OF LYSOZYME

EC 3·2·1·17, Lysozyme (muramidase, mucopeptide *N*-acetylhydrolase) from hen's egg white is a polypeptide of 129 amino acid residues having a molecular weight of 14,400 dalton. It is a basic protein with an isoelectric point of 10.7-11.0. In the hen's egg white, lysozyme accounts for 3.5% of the total egg white protein. Lysozyme is an enzyme which has the ability to lyse certain gram positive bacteria by hydrolyzing the β-linkage between *N*-acetylmuramic acid (NAM) and *N*-acetylglucosamine (NAG) of mucopolysaccharides in the peptidoglycan of the bacterial cell wall, as shown in Figure 6. The structural information and enzymatic behavior of lysozyme are clearly defined and have been summarized in a book entitled *Lysozyme* [51].

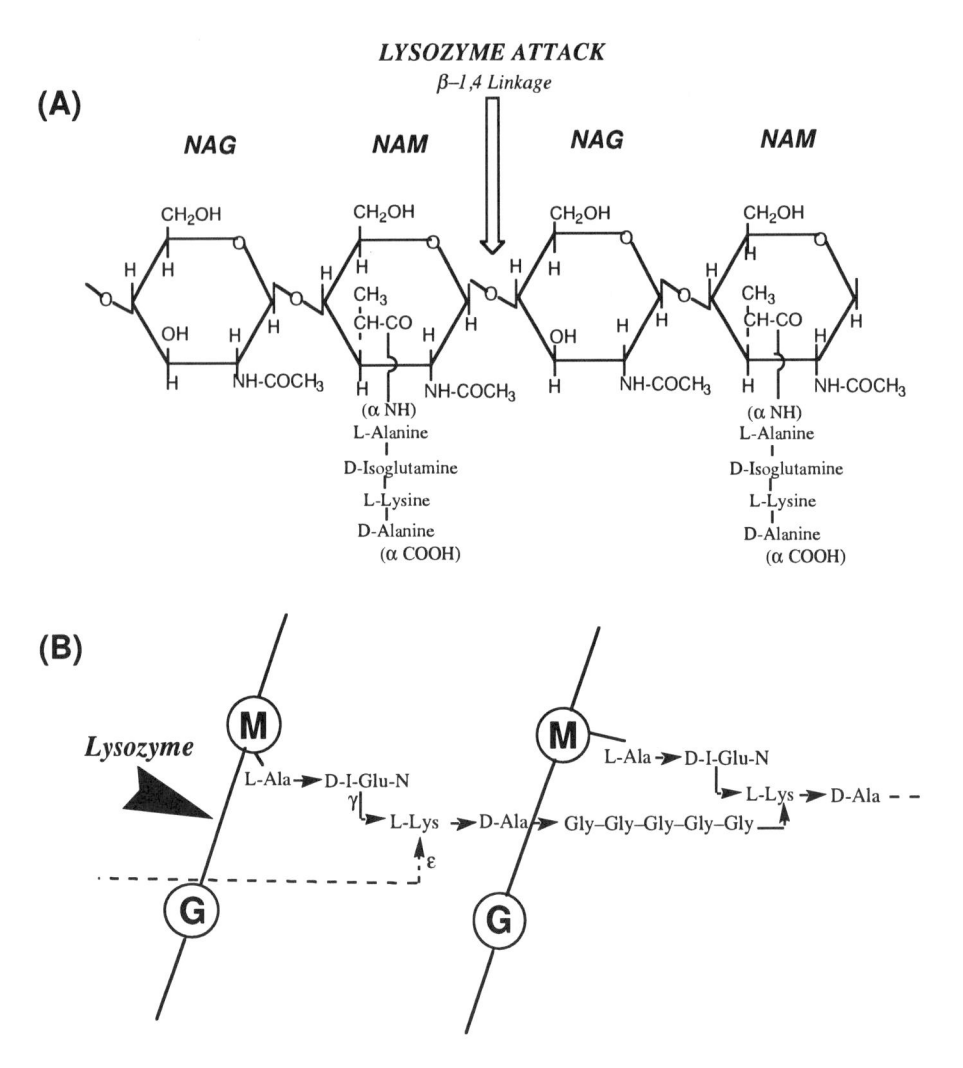

Figure 6. Primary structure of the peptidoglycan in the wall of *Staphylococcus aureus* showing the cutting site of lysozyme. (A) A portion of the glycan strand of the peptidoglycan where lysozyme attacks the glycosidic bond between N-acetylglucosamine (NAG) and N-acetylmuramic acid (NAM). The COOH of the D-lactyl groups of muramic acid is usually peptide substituted. (B) Schematic presentation of the peptidoglycan lattice that is cross-linked through pentaglycine peptide chains.

Lysozyme is the first protein which was sequenced [52] and whose three-dimensional structure was completely analyzed [53]. Hen lysozyme is a monomeric, secretory protein containing four disulfide bonds: Cys 6-127, 30-115, 64-80, and 76-94, with no free sulfhydryl groups, and its tertiary structure is considerably stable in aqueous solution. When its disulfide bonds are lost, its native conformation transforms to a loose structure. The lysozyme molecule is ovoid and consists of two domains or lobes, linked by a long α–helix between which lies the active site of the enzyme. The lower N-terminal lobe (residues 40-88) consists of some helices and is mostly an antiparallel β-sheet. The second lobe is made up of residues 1-39 and 89-129, and its secondary structure is largely α–helical [54]. The two lobes of the molecule are linked via interaction between Lys 97 and Phe 38, as demonstrated in Figure 7. The molecule conforms to the principle of hydrophobic in, hydrophilic out of a protein. All of its polar groups are on the surface and the majority of nonpolar (hydrophobic) groups are buried in the interior. Conformational transition in lysozyme involves the relative movement of its two lobes to each other in a cooperative manner, thus, allowing significant movement of structural regions within the folding unit of the molecule. This new concept has been referred to as "hinge-bending." A schematic representation of one concept of hinge-bending is shown in Figure 8. As can be seen, the relative movement of the domains of the lysozyme molecule can cause large global conformational changes which may permit access to the substrate and generate an appropriate environment for catalysis. The conformational flexibility offered by the interaction between Lys 97 and Phe 38 (Figure 7) is of particular note in that it may be important in the formation of a hinge-bending structure which is thought to play a critical role in the enzymatic action.

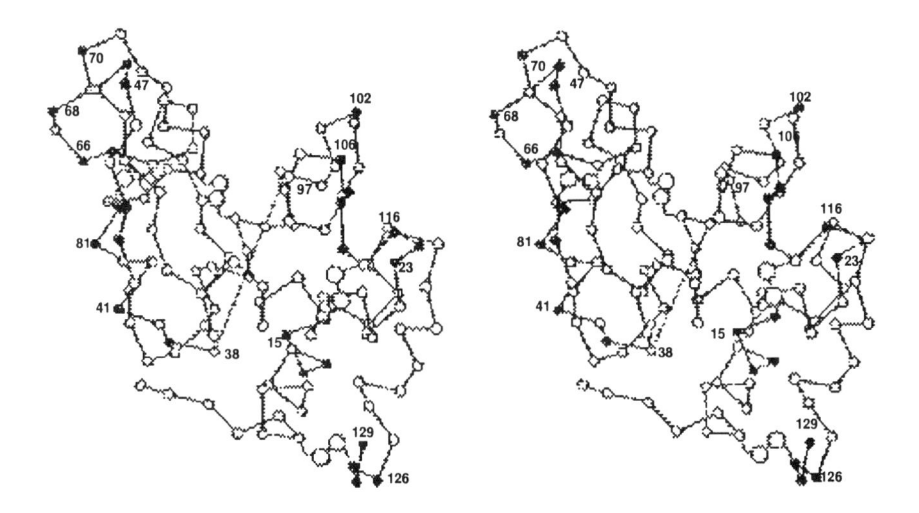

Figure 7. Stereo view of the C^α chain of the hen egg white lysozyme molecule. Interaction between $C^\alpha 38$ and $C^\alpha 97$ defining the hinge is represented by a dashed line. Disulfide bonds are represented by large circles. Residues marked with filled circles are those involved in intermolecular contacts. Reproduced from Young et al. (1994) [54] by permission of Academic Press, Ltd., London.

The active site of hen egg white lysozyme consists of six subsites A, B, C, D, E, and F which together are sufficient for binding six sugar residues [55]. The six subsites along the active site cleft positioning the catalytic groups Glu-35 and Asp-52 between subsites D and E.

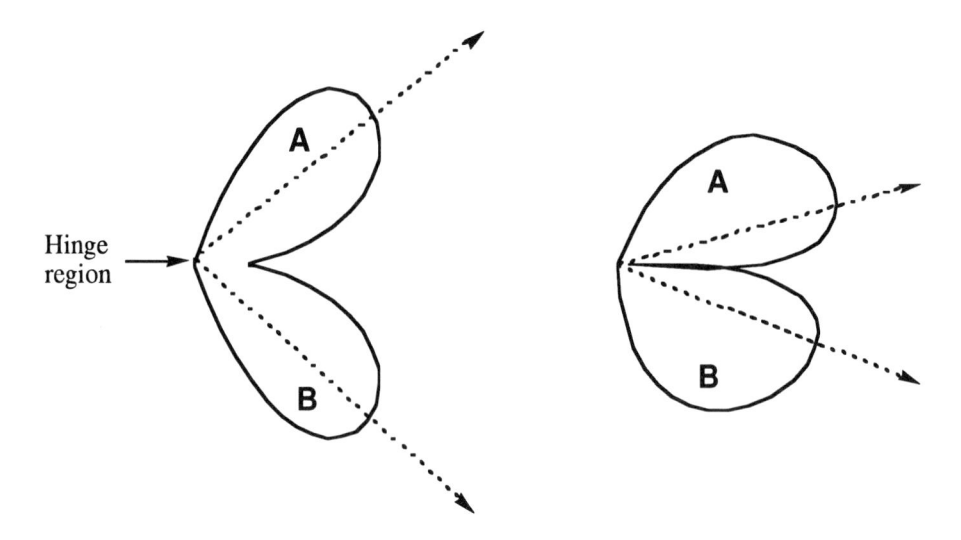

Figure 8. A schematic representation of the hinge-bending structure of lysozyme. The two lobes of lysozyme (A and B) are linked together by a segment. The relative movement of the two lobes in this way can cause large global conformational changes.

B. BIOLOGICAL FUNCTIONS OF LYSOZYME

In spite of the huge number of biochemical investigations devoted to lysozyme, its role in various physiological systems (tissues) including hen's egg white is still unclear. Regardless of the direct bacteriolytic action [56], many other biological functions of lysozyme have recently been reported, such as antiviral action to inactivate certain viruses by forming an insoluble complex with acidic viruses [57], potentiation of antibiotic effects [58], anti-inflammatory action [51], direct activation of immune cells (monocytes and lymphocytes) [59], antitumor action [19], fusogenic activity to phospholipids [60], and agglutinating and antiheparinic action [61].

The antimicrobial activity of lysozyme is limited to certain gram positive bacteria. The gram negative bacteria are less susceptible to the bacteriolytic action of the enzyme. The susceptibility differences are believed to be due to the complex envelope structure of the latter. Gram-negative bacteria such as *E. coli* or *Salmonella typhimurium* are covered by a double membrane complex. The cytoplasmic membrane is surrounded by the outer membrane and the peptidoglycan layer (the substrate of lysozyme) is located between these two membranes. Therefore, the outer membrane serves as a barrier to the access of lysozyme to its site of action (peptidoglycan). Disruption of the outer membrane integrity (usually by EDTA or proteinases) usually renders the bacteria sensitive to the lysozyme action [62, 63]. The lower sensitivity of gram negative bacteria to lysozyme may not be attributed merely to the inaccessibility of the peptidoglycan layer to the enzyme, because lysozyme is found capable of distorting the outer membrane, releasing intracellular proteins without dramatic effect on the bacterial cellular viability [64]. One possible explanation can be that lysozyme is entrapped in the outer membrane through specific binding to the lipopolysaccharides (LPS) contained and, hence, becomes arrested at this site. The specific binding and inactivation of lysozyme by LPS has been reported [16, 66]. In accordance with this hypothesis, we were able to demonstrate that structural modification of lysozyme with a membrane fusing hydrophobic domain, such as saturated fatty acids or hydrophobic peptide, can enhance the insertion of lysozyme into the bacterial membrane and the subsequent killing of gram negative bacteria [64, 66]. From a different point of view, the limited action of lysozyme on gram negative bacteria seems to be influenced by the composition and the sequence of the *N*-acetylamino sugars of the bacterial cell walls. In this regard, the isolated cell walls of different bacteria showed varying degrees of susceptibility to digestion with hen egg white lysozyme [67]. Of many different gram positive bacteria, the walls of *Micrococcus lysodeikticus* were the most sensitive and the walls of *Staphylococci* were less sensitive to the bacteriolytic action of lysozyme. Among gram negative bacteria, the walls of *Salmonella* and *Shigella* were the most sensitive whereas that of *E. coli, Vibrio,* and *Proteus* were very much less sensitive [68]. In addition to the above data, there was evidence that lysozyme devoid of enzymatic activity kills different types of bacteria. Therefore, it should be warned that the bacteriostatic, bacteriolytic, and bactericidal activities of lysozyme seem to operate by different mechanisms.

Concerning immunostimulatory activity, binding of lysozyme to the bacterial endotoxic LPS is thought to play an important role in the subsequent phagocytosis of bacteria invading the body. The high affinity binding of lysozyme from egg white to bacterial LPS has been demonstrated to result in reversible inactivation of its enzymatic activity [16, 65]. The major binding sites of LPS on lysozyme were found to be two negatively charged, hydrophobic regions responsible for hydrophobic interactions. It was documented that release of LPS from the outer membrane of bacteria usually followed by interaction with immune cells, particularly macrophages, can stimulate these cells. The lysozyme-LPS complex was suggested to contribute to the capacity of the host to respond to infection by gram negative bacteria [65]. In another study, it has been revealed that lysosomal proteinases of azurophil granules, such as elastase and cathepsin B, D, and G, rendered the gram negative bacteria sensitive to lysozyme [69]. It is known that upon phagocytosis of bacteria, enzymes from the inflammatory cells, polymorphonuclear leukocytes, are discharged into the phagocytic vacuole where killing and digestion subsequently occur. The lysosomal azurophil granules contain several proteinases, such as elastase and cathepsins, in addition to the abundant lysozyme content. It has been concluded that, in the presence of lysozyme, lysosomal proteinases are probably of only secondary importance, allowing further degradation of cell wall fragments after the action of lysozyme. They also observed the concerted attack of lysosomal lysozyme and proteinases in morphologic studies with rabbit polymorphonuclear leukocytes. From the above studies showing that lysozyme has different binding capacities to either polysaccharides or lipids of bacteria, it is clear that lysozyme possesses direct and indirect antimicrobial actions.

We have recently reported a series of interesting approaches to prove that lysozyme can be lethal to gram negative bacteria if its interaction with the bacterial membrane was strengthened by modifying its surface hydrophobicity. Hen's egg white lysozyme was chemically modified with palmitate or stearate residues [62, 66] or genetically fused with a hydrophobic pentapeptide of the same length as the palmitic acid residue [64]. All of the lysozyme derivatives exhibited strong bactericidal action to *E. coli* k 12. They were capable of inserting into the lipid bilayer and subsequently disrupting the electrochemical potential by forming ion pores in the membrane. The data suggested that the lysozyme molecule can be intrinsically lethal to bacteria if its structure supplies a hydrophobic domain at the surface of a molecule. However, it will be interesting to see whether the conformational transition in lysozyme can account for its variable antimicrobial action or not.

C. CONFORMATIONAL TRANSITION OF LYSOZYME

Now we have come to a very fundamental point. Lysozyme was found to exist in two conformational states between 20 and 30°C with a transition point at 25°C [15]. The pH-dependent conformational transition of lysozyme has also been reported [70]. A decrease in the affinity of lysozyme for its strong inhibitor N-acetylglucosamine was noted at a neutral pH above the transition temperature (25°C). Investigations of the action of this amino sugar at 40°C (the physiological temperature of birds) were attempted at various pHs and the inhibitor-insensitive lysozyme forms were characterized [71]. These studies suggested the existence of two different temperature-induced rearrangements in which lysozyme behaves differently, presumably because of different conformations. These were called the A and B forms (low and high temperature forms, respectively).

Lysozyme is normally present as a reversible dimer between pH 5.0 and 9.0. Hen's egg white lysozyme tends to associate in an irreversible dimeric form when hen eggs are stored for long periods. The dimer retains the enzymatic activity of the monomer, so this reaction is not responsible for loss of activity. This dimerization and, in some instances higher polymerization depends on pH, concentration, and temperature [70]. In this regard, a recent study on a site-specific dimerized lysozyme, achieved by the intermolecular cross-linking of the mutant monomer that contained either Arg 41 to Cys or Ala 73 to Cys substitution with a divalent maleimide, exhibited altered substrate specificity [72], as shown in Figure 9. The dimer R41C-R41C having the two catalytic clefts close to each other was 2.3 times more specific to a polymer substrate, ethylene glycol chitin, compared to an oligomeric substrate, (GlcNAc) [5]. The R41C-R41C dimer showed approximately half the relative activity against an oligomeric substrate (38%) of that of the other dimers, R41C-A73C and A73C-A73C (97% and 90%).

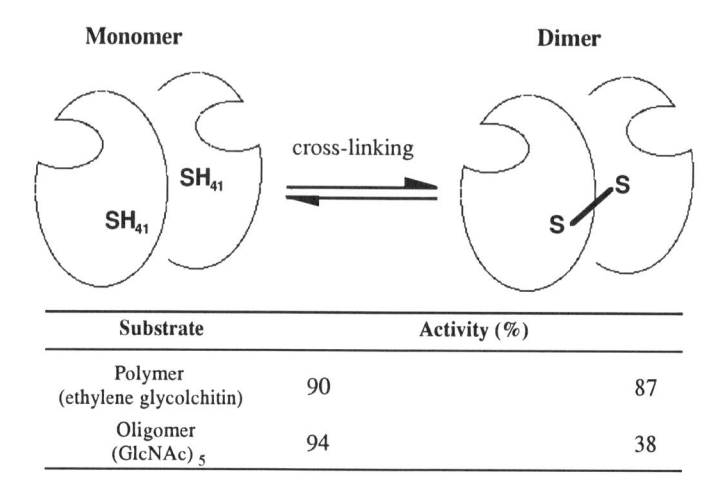

Substrate	Activity (%)	
Polymer (ethylene glycolchitin)	90	87
Oligomer (GlcNAc)$_5$	94	38

Figure 9. Dimerization of mutant lysozyme containing Cys at residue 41, shifts the substrate specificity to the polymeric substrate, likely by altering the binding mode of the substrate to the active site. Adapted from Muraki et al. (1994) [72] by permission of Elsevier Science Publishers B. V., Amsterdam.

The results lead to the speculation that the dimerization of lysozyme in the hen egg white may operationally mimic the catalytic behavior of the artificially dimerized lysozyme in the above mentioned approach. It will be of general interest to see whether other techniques can provide more direct information about the state of substrate specificity of the naturally dimerized lysozyme in egg white. Yet another principle emerging from the structural data on lysozyme is that the active site of the enzyme is located in a region of high displacement from which the globular intramolecular motion has been suggested. In line with this, a specific temperature- and pH-dependent conformational transition was observed in lysozyme. This globular mobility of the lysozyme may play a significant role in its variable biological functions, particularly of its A and B forms.

D. THERMAL STABILITY AND DENATURATION

Since lysozyme has four disulfide bonds making the molecule of 129 amino acid residues unusually compact, its heat stability would be considerably high. Lysozyme is known to be stable at 100°C for 2 min at pH 4.5. It is rapidly inactivated by thiol compounds. In hen egg white the activity of lysozyme does not survive when heated at 60°C because of the presence of the free sulfhydryl groups of ovalbumin. The chemical reactions to account for the irreversible denaturation of hen egg white lysozyme have been studied in recent years [73, 74]. The studies related the irreversible inactivation of lysozyme at pH 4.0 and 100°C to the accumulation of multiple chemical reactions, such as isomerization of the Asp-Gly sequence, deamidation of the Asn and Gln residues, racemization of the Asp and Asn residues, and cleavage of the Asp-X peptide bond. The other study devoted to the deamidation of lysozyme obtained several important observations [74]. A new peak on a cation-exchange column appeared with increasing incubation time of lysozyme at pH 4.0 and 40°C up to 10 days. The derivative was identified as 101-succinimide lysozyme where the cyclic imide formed at Asp 101-Gly 102. The lytic activity of the 101-succinimide lysozyme was 125% of the native lysozyme. The cyclic imide was completely converted into the β-linkage by further incubation at pH 8.0. The lytic activity of the β-lysozyme was 165% of the native lysozyme. The mechanism of the conversion of α–Asp-Gly sequence into the β-form via formation of the cyclic imide has also been proposed, as shown in Scheme 1. The conclusion was that formation of the cyclic imide requires accessibility of the side chain at the surface of a molecule and flexibility of the main chain. Interestingly, deamidation of lysozyme during the storage of hen eggs has recently been reported [75], and it was found as enzymatically active as the fresh lysozyme preparation.

Indeed, more information about what happens to lysozyme in stored or fertilized eggs may reveal the actual multiple physiological roles of lysozyme in intact eggs.

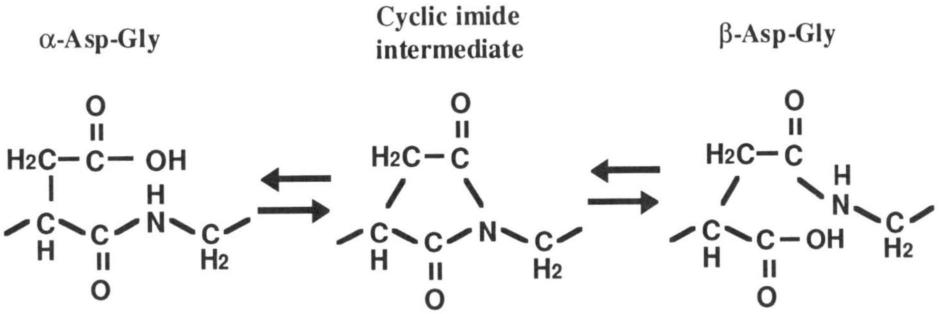

Scheme 1. Isomerization of the Asp-Gly sequence in heat-treated lysozyme. Reproduced from Tomizawa et al. (1994) [74] by permission of American Chemical Society, Washington, D. C.

V. CONCLUDING REMARKS

The occurrence of multiple conformational states of proteins in the egg is expected to have important consequences for interpretation of their biological roles. Recent advances in studies of protein dynamics support the picture of a rich variety of structurally similar conformations that interconvert rapidly at physiological temperatures. During embryogenesis of an egg, diversified inert protein structures interact in a certain fashion via a chain of stages to produce the living chick. These stages are initiated just by fertilization and periodical incubation of the egg at 37-40°C. It appears that these stages involve several complexes through biological interconversion of the constituent proteins. The solids of egg white consist mainly of proteins (11%), having a complex of multiple structural and functional properties. They can be simply classified on the basis of their functions as follows: (1) proteinase-inhibitors, (2) antimicrobial, (3) metal and vitamins chelators, (4) jelly-like structure. The multiple roles of this set of proteins are believed to prevent the penetration of invading microorganisms from reaching the yolk and the embryo. However, egg white is not in direct contact with the embryo and it is consumed very slowly during embryogenesis. During incubation, egg white shrinks in size as the water and solutes pass into the yolk. An early study suggested that about 30 % of egg white proteins are adsorbed through the yolk sac during the last week of incubation [76]. Dephosphorylated ovalbumin was also found in embryonic blood serum after two weeks of incubation. Pons and his colleagues found a gradual movement of free amino acids from the yolk to the white of fertile eggs during the first day of incubation [77]. The binding of the free amino acids to egg white proteins has also been detected. The picture which emerged from the present article and the above mentioned literature is that the induction of embryogenesis seems to be accompanied by changes in the functional properties of egg white proteins as a consequence of some allosteric activation or inhibition. Dehydration as the water diffuses to the yolk, binding of free amino acids, phosphorylation, proteolytic cleavage, or deglycosylation of egg white proteins may take a place in such activation. The structural similarity between ovalbumin, cleaved with egg white proteinases and other proteins of various biological functions would provide interesting information about the actual role of ovalbumin in the egg. The multiple roles of lysozyme uncovered so far deny that lysozyme still has unknown function(s) in the egg or during embryogenesis. This assumption can be confirmed if some allosteric or proteolytic activation were detected in the egg white. This should await new techniques to study protein structure which can be applied *in situ* or under conditions that are relevant to their actual environment.

REFERENCES

1. **Vadehra, D. V. and Nath, K. R.**, Eggs as a source of proteins, *CRC Crit. Rev. Food Tech.*, **4**, 193, 1973.
2. **Tabe, L., Krieg, P., Strachan, R., Jackson, D., Wallis, E., and Colman, A.**, Segregation of mutant ovalbumins and ovalbumin-globin fusion proteins in *Xenopus* oocytes. Identification of an ovalbumin signal sequence, *J. Mol. Biol.*, **180**, 645, 1984.
3. **Fidelio, G. D., Austen, B. M., Chapman, D., and Lucy, J. A.**, Interactions of ovalbumin and of its putative signal sequence with phospholipid monolayers. Possible importance of differing lateral stabilities in protein translocation, *Biochem. J.*, **244**, 295, 1987.
4. **Yun, C. H. and Kim, H.**, Ovalbumin-induced fusion of acidic phospholipid vesicles at

low pH, *J. Biochem.*, **105**, 406, 1989.
5. **Yun, T., Yamashita, H., Takahashi, N., and Hirose, M.**, Limited proteolysis of disulfide-reduced ovalbumin by subtilisin, *Biosci. Biotech. Biochem.*, **57**, 940, 1993.
6. **Wright, H. T., Xi-Qian, H. and Huber, R.**, Crystal structure of plakalbumin, a proteolytically nicked form of ovalbumin, *J. Mol. Biol.*, **213**, 513, 1990.
7. **Jeltsch, J. M. and Chambon, P.**, The complete nucleotide sequence of the chicken ovotransferrin mRNA, *Eur. J. Biochem.*, **122**, 291, 1982.
8. **Oratore, A., D'Andrea, G., Moreton, K., and Williams, J.**, Binding of various ovotransferrin fragments to chick-embryo red cells, *Biochem. J.*, **257**, 301, 1989.
9. **Feeney, R. E., Stevens, F. C., and Osuga, D. T.**, The specificities of chicken ovomucoid and ovoinhibitor, *J. Biol. Chem.*, **238**, 1415, 1963.
10. **Rhodes, M. B., Bennett, N., and Feeney, R. E.**, The trypsin and chymotrypsin Inhibitors from avian egg whites, *J. Biol. Chem.*, **235**, 1686, 1960.
11. **Haynes, R. and Feeney, R. E.**, Fractionation and properties of trypsin and chymotrypsin inhibitors from lima beans, *J. Biol. Chem.*, **242**, 5378, 1967.
12. **Garibaldi, J. A.**, Factors in egg white which control growth of bacteria, *Food Res.*, **25**, 337, 1960.
13. **Schechter, Y., Burstein, Y., and Gertler, A.**, Effect of oxidation of methionine residues in chicken ovoinhibitor on its inhibitory activities against trypsin, chymotrypsin, and elastase, *Biochemistry*, **16**, 992, 1977.
14. **Gertler, A. and Feinstein, G.**, Inhibition of porcine elastase by turkey ovomucoid and chicken ovoinhibitor, *Eur. J. Biochem.*, **20**, 547, 1971.
15. **Jolles, P. and Jolles, J.**, What's new in lysozyme research?, *Mol. Cell. Biochem.*, **63**, 165, 1984.
16. **Ohno, N. and Morrison, D. C.**, Lipopolysaccharide interaction with lysozyme. Binding of lipopolysaccharide to lysozyme and inhibition of lysozyme enzymatic activity, *J. Biol. Chem.*, **264**, 4434, 1989.
17. **Hayakawa, S., Kondo, H., Nakamura, R., and Sato, Y.**, Effect of β-ovomucin on the solubility of α-ovomucin and further inspection of the structure of ovomucin complex in thick egg white, *Agric. Biol. Chem.*, **47**, 815, 1983.
18. **Oliver, C. N. and Stadtman, E. R.**, A proteolytic artifact associated with the lysis of bacteria by egg white lysozyme, *Proc. Natl. Acad. Sci. USA*, **80**, 2156, 1983.
19. **Sava, G.**, Reduction of B16 melanoma metastases by oral administration of eggwhite lysozyme, *Cancer Chemother. Pharmacol.*, **25**, 221, 1989.
20. **Keilova, H. and Tomasek, V.**, Effect of papain inhibitor from chicken egg white on cathepsin B_1, *Biochim. Biophys. Acta*, **334**, 179, 1974.
21. **Green, N. M.**, Avidin, *Adv. Protein Chem.*, **29**, 85, 1975.
22. **Murthy, U. S., Podder, S. K., and Adiga, P. R.**, The interaction of riboflavin with a protein isolated from hen's egg white: A spectrofluorimetric study, *Biochim. Biophys. Acta*, **434**, 69, 1976.
23. **Nagase, H., Harris, E. D., Jr., Woessner, J. F., Jr., and Brew, K.**, Ovostatin: A novel proteinase inhibitor from chiken egg white, *J. Biol. Chem.*, **258**, 7481, 1983.
24. **Nagase, H., Harris, E. D., Jr., and Brew, K.**, Evidence for a thiol ester in duck ovostatin (ovomacroglobulin), *J. Biol. Chem.*, **261**, 1421, 1986.
25. **Petrovic, S. and Vitale, L.**, Purification and properties of glutamyl aminopeptidase from chicken eggwhite, *Comp. Biochem. Physiol.*, **95B**, 589, 1990.
26. **Donovan, J. W. and Hansen, L. U.**, The β-acetylglucosaminidase activity of eggwhite. 1. Kinetics of the reaction and determination of the factors affecting the stability of the enzyme in egg white, *J. Food Sci.*, **36**, 174, 1971.

27. **Burley, R. W. and Vadehra, D. V.,** The Albumen: Chemistry, in *The Avian Egg. Chemistry and Biology*, John Wiley & Sons, Inc., New York, 1989.
28. **Woo, S. L. C., Beattie, W. G., Catterall, J. F., Dugaiczyk, A., Staden, R., Brownlee, G. G., and O'Malley, B. W.,** Complete nucleotide sequence of the chicken chromosomal ovalbumin gene and its biological significance, *Biochemistry*, **20**, 6437, 1981.
29. **Nisbet, A. D., Saundry, R. H., Moir, A. J. G., Fothergill, L. A., and Fothergill, J. E.,** The complete amino acid sequence of hen ovalbumin, *Eur. J. Biochem.*, **115**, 335, 1981.
30. **Thompson, E. O. P. and Fisher, W. K.,** Amino acid sequences containing half-cystine residues in ovalbumin, *Aust. J. Biol. Sci.*, **31**, 433, 1978.
31. **Smith, M. B.,** Studies on ovalbumin I. Denaturation by heat and the heterogeneity of ovalbumin, *Aust. J. Biol. Sci.*, **17**, 261, 1964.
32. **Atkinson, P. H., Grey, A., Carver, J. P., Hakimi, J., and Ceccarini, C.,** Demonstration of heterogeneity of chick ovalbumin glycopeptides using 360-MHz proton magnetic resonance spectroscopy, *Biochemistry*, **20**, 3979, 1981.
33. **Stein, P. E., Leslie, A. G. W., Finch, J. T., and Carrell, R. W.,** Crystal structure of uncleaved ovalbumin at 1.95 Å resolution, *J. Mol. Biol.*, **221**, 941, 1991.
34. **Satake, K., Sasakawa, S., and Honda, S.,** Formation of plakalbumin from ovalbumin under the action of plant proteinases, *J. Biochem.*, **58**, 305, 1965.
35. **Carrell, R. W. and Travis, J.,** α1-antitrypsin and the serpins: Variation and countervariation, *Trends Biochem. Sci.*, **10**, 20, 1985.
36. **Carrell, R. W., Pemberton, P., and Boswell, D. R.,** The serpins: Evolution and adaptation in a family of protease inhibitors, *Cold Spring Harbor Symp. Quant. Biol.*, **52**, 527, 1987.
37. **Debruyne, I. and Stockx, J.,** The nucleoside triphosphate and nucleoside diphosphate phosphorylase, the 5'-nucleotidase and alkaline phosphatase activities in the hen's egg yolk, *Arch. Int. Physiol. Biochem.*, **84**, 148, 1976.
38. **Kato, A. and Takagi, T.,** Formation of intermolecular β-sheet structure during heat denaturation of ovalbumin, *J. Agric. Food Chem.*, **36**, 1156, 1988.
39. **Aisen, P. and Listowsky, I.,** Iron transport and storage proteins, *Annu. Rev. Biochem.*, **49**, 357, 1980.
40. **Brock, J. H.,** Transferrins, in *Metalloproteins, Part 2: Metal proteins with non-redox roles,* Harrison, P. M., Ed., Verlag Chemie, GmbH, Weinheim, 1985.
41. **Metz-Bovtigue, M. H., Jolles, J., Mazurier, J., Schoentgen, F., Legrand, D., Spik, G., Montreuil, J., and Jolles, P.,** Human lactotransferrin: Amino acid sequence and structural comparisons with other transferrins, *Eur. J. Biochem.*, **145**, 659, 1984.
42. **Pierce, A., Colavizza, D., Benaissa, M., maes, P., Tartar, A., Montreuil, J., and Spik, G.,** Molecular cloning and sequence analysis of bovine lactotransferrin, *Eur. J. Biochem.*, **196**, 177, 1991.
43. **Williams, J., Elleman, T. C., Kingston, I. B., wilkins, A. G., and Kuhn, K. A.,** The primary structure of hen ovotransferrin, *Eur. J. Biochem.*, **122**, 297, 1982.
44. **Baker, E. N., Rumball, S. V., and Anderson, B. F.,** Transferrins: Insights into structure and function from studies on lactoferrin, *Trends Biochem. Sci.*, **12**, 350, 1987.
45. **Ikeda, H., Banuki, Y., Nakazato, K., Tanaka, Y., and Satake, K.,** Preparation and characterization of trypsin-nicked ovotransferrin, *FEBS Lett.*, **182**, 305, 1985.
46. **Brown-Mason, A., Brown, S., Butcher, N. D. and Woodworth, R. C.,** Reversible association of half-molecules of ovotransferrin in solution. Basis of cooperative binding to reticulocytes, *Biochem. J.*, **245**, 103, 1987.
47. **Schade, A. L. and Caroline, L.,** Raw hen egg white and the role of iron in growth inhibition of *Shigella dysenteriae, Staphylococcus aureus, Escherichia coli* and *Saccharomyces cerevisiae, Science*, **100**, 14, 1944.

48. **Valenti, P., Antonin, C., Von Hunolstein, P., Visca, P., Orsi, N., and Antonini, E.,** Studies of the antimicrobial activity of ovotransferrin, *Int. J. Tissue React.*, **1**, 97, 1983.
49. **Boesman-Finkelstein, M. and Finkelstein, R.,** Antimicrobial effects of human milk: Inhibitory activity on enteric pathogens, *FEMS Microbiol. Immunol.*, **27**, 167, 1985.
50. **Bellamy, W., Takase, M., Yamauchi, K., Wakabayashi, H., Kawase, K., and Tomita, M.,** Identification of the bactericidal domain of lactoferrin, *Biochim. Biophys. Acta*, **1121**, 130, 1992.
51. **Osserman, E. F., Canfield, R. E., and Beychok, S.,** *Lysozyme*, Osserman, E. F., Canfield, R. E., and Beychok, S., Eds., Academic Press, New York and London, 1974.
52. **Blake, C. C. F., Koenig, D. F., Mair, G. A., North, A. C. T., Phillips, D. C., and Sarma, V. R.,** Structure of hen egg-white lysozyme, *Nature*, **206**, 757, 1965.
53. **Diamond, R.,** Real-space refinement of the structure of hen egg-white lysozyme, *J. Mol. Biol.*, **82**, 371, 1974.
54. **Young, A. C. M., Tilton, R. F., and Dewan, J. C.,** Thermal expansion of hen egg-white lysozyme. Comparison of The 1.9 Å resolution structures of the tetragonal form of the enzyme at 100 K and 298 K, *J. Mol. Biol.*, **235**, 302, 1994.
55. **Perkins, S. J., Johnson, L. N., Phillips, D. C., and Dwek, R. A.,** The binding of monosaccharide inhibitors to hen egg-white lysozyme by proton magnetic resonance at 270 MHz and analysis by ring-current calculations, *Biochem. J.*, **193**, 553, 1981.
56. **Bigger, W. D. and Sturgess, J. M.,** Role of lysozyme in the microbicidal activity of rat alveolar macrophages, *Infect. Immun.*, **16**, 974, 1977.
57. **Hasselberger, F. X.,** *Uses of Enzymes and Immobilized Enzymes*, Nelson-Hall Inc., Chicago, 1978.
58. **El-Nimr, A., Hardee, G. E., and Perrin, J. H.,** A fluorimetric investigation of the binding of drugs to lysozyme, *J. Pharm. Pharmacol.*, **33**, 117, 1981.
59. **Rinehart, J., Jacobs, H., and Osserman, E.,** Lysozyme modulation of lymphocyte proliferation, *Clin. Res.*, **27**, 305, 1979.
60. **Posse, E., Dearcuri, B. F., and Morero, R. D.,** Lysozyme interactions with phospholipid vesicles: Relationships with fusion and release of aqueous content, *Biochim. Biophys. Acta-Biomembranes*, **1193**, 101, 1994.
61. **Mega, T. and Hase, S.,** Conversion of egg-white lysozyme to a lectin-like protein with agglutinating activity analogous to wheat germ agglutinin, *Biochim. Biophys. Acta-Gen. Subjects*, **1200**, 331, 1994.
62. **Ibrahim, H. R., Kato, A., and Kobayashi, K.,** Antimicrobial effects of lysozyme against gram-negative bacteria due to covalent binding of palmitic acid, *J. Agric. Food Chem.*, **39**, 2077, 1991.
63. **Thorne, K. J. I., Oliver, R. C., and Barrett, A. J.,** Lysis and killing of bacteria by lysosomal proteinases, *Infect. Immun.*, **14**, 555, 1976.
64. **Ibrahim, H. R., Yamada, M., Matsushita, K., Kobayashi, K., and Kato, A.,** Enhanced bactericidal action of lysozyme to *Escherichia coli* by inserting a hydrophobic pentapeptide into C Terminus, *J. Biol. Chem.*, **269**, 5059, 1994.
65. **Ohno, N. and Morrison, D. C.,** Effects of lipopolysaccharide chemotype structure on binding and inactivation of hen egg lysozyme, *Eur. J. Biochem.*, **186**, 621, 1989.
66. **Ibrahim, H. R., Kobayashi, K., and Kato, A.,** Length of hydrocarbon chain and antimicrobial action to gram-negative bacteria of fatty acylated lysozyme, *J. Agric. Food Chem.*, **41**, 1164, 1993.
67. **Salton, M. R. J. and Pavlik, J. G.,** Studies of the bacterial cell wall VI. Wall composition and sensitivity to lysozyme, *Biochim. Biophys. Acta*, **39**, 398, 1960.
68. **Peterson, R. G. and Hartsell, S. E.,** The lysozyme spectrum of the gram-negative bacteria,

J. Infect. Dis., **96**, 75, 1955.
69. **Thorne, K. J., Oliver, R. C., and Barret, A. J.**, Lysis and killing of bacteria by lysosomal proteinases, *Infect. Immun.*, **14**, 555, 1976.
70. **Sophianopoulos, A. J. and Holde, K. E.**, Physical studies of muramidase (lysozyme), *J. Biol. Chem.*, **239**, 2516, 1964.
71. **Saint-Blanced, J., Mazurier, J., Bournaud, M., Maurel, P. P., Berthou, J., and Jolles, P.**, The temperature and pH-dependent transition of hen lysozyme. Characterization of two temperature-defined domains and of an *N*-acetylglucosamine (inhibitor)- insensitive form, *Mol. Biol. Rep.*, **5**, 165, 1979.
72. **Muraki, M., Jigami, Y., and Harata, K.**, Alteration of the substrate specificity of human lysozyme by site-specific intermolecular cross-linking, *FEBS Letters*, **355**, 271, 1994.
73. **Tomizawa, H., Yamada, H., and Imoto, T.**, The mechanism of irreversible inactivation of lysozyme at pH 4 and 100°C, *Biochemistry*, **33**, 13032, 1994.
74. **Tomizawa, H., Yamada, H., Ueda, T., and Imoto, T.**, Isolation and characterization of 101-succinimide lysozyme that possesses the cyclic imide at Asp101-Gly102, *Biochemistry*, **33**, 8770, 1994.
75. **Kato, A., Shibata, M., Yamaoka, H., and Kobayashi, K.**, Deamidation of lysozyme during the storage of egg white, *Agric. Biol. Chem.*, **52**, 1973, 1988.
76. **Saito, Z. and Martin, W. G.**, Ovalbumin and other water-soluble proteins in avian yolk during embryogenesis, *Can. J. Biochem.*, **44**, 293, 1966.
77. **Pons, A., Garcia, F. J., Palou, A., and Alemany, M.**, Permeability of chicken egg vitelline membrane to amino acids-binding of amino acids to egg proteins, *Comp. Biochem. Physiol.*, **82A**, 289, 1985.

Chapter 5

EGG YOLK PROTEINS

L.R. Juneja and M. Kim

TABLE OF CONTENTS

I. Introduction
II. Classes of Egg Yolk Protein
    A. Lipovitellin (High density lipoproteins)
    B. Lipovitellenin (Low density lipoproteins)
    C. Phosvitin
    D. Livetin
III. Characteristic Structures of Egg Yolk Proteins
IV. Emulsifying Properties of and Egg Yolk Proteins
V. Gelling Properties of Egg Yolk Proteins
VI. Heat Stability of Egg Yolk Proteins
VII. Commercial Products of Egg Yolk Proteins
    A. Preparation of EYP-80
    B. Amino Acid Pattern of EYP-80
    C. Minerals and Vitamins of EYP-80
    D. Antioxidant Activity of EYP-80
    E. Cholesterol Content of EYP-80
VIII. Concluding Remarks

References

I. INTRODUCTION

The word "protein" originated from the Greek verb meaning to "take place first." Proteins as a nutritive food material are ever fashionable and are obtained from various biological sources. The importance of proteins to overall diet and health has been recognized for many years; however, the quality of protein sources has been the subject of great debate.

A protein is utilized only in so far as it fulfills the minimal requirement for each of the essential amino acids. If one of them is absent entirely in proteins in food and not furnished by any other supplement, there is an immediate loss of nitrogen from the body and this, of course, cannot be made up even by a larger intake of the same deficient protein. The organisms minimal requirements for the amino acid that are relatively lacking can be met if consumption of the protein is sufficiently large. A protein may be tested for amino acid balance and general excellence by determining the smallest amount of amino acid is contains.

Compared with the hen egg, no other single food of animal origin is eaten and relished by so many people the world over [1]. Eggs are popular not only due to their availability but also because of excellent nutritional value. The hen egg, viewed by many as one of nature's most perfect foods, has been a staple food for humans since the dawn of civilization.

Egg yolk is the largest biological cell known which originates from one cell division [1] and is composed of various important chemical substances that form the basis of life. Therefore, an avian egg is a storehouse of nutrients, for example, proteins, lipids, carbohydrates, and biologically active substances including growth promoting factors which are needed to form a chicken. Protein is essential for building the components of life. Hen eggs provide the most complete, therefore, the highest quality proteins with all necessary vitamins and minerals. Man's utilization of egg proteins is likewise excellent, as indicated by the high biological value of eggs.

In supporting the most exacting of life processes, reproduction, lactation, and growth, the proteins of the egg, when constituting 20 or 30% of the diet, are better than those of milk and muscle meats both of which are considered good sources of complete proteins. When eggs are fed to rats as 10 or 15% of the diet, the growth-promoting value of whole egg is superior to that of wheat germ, corn germ, soybean flour, cottonseed flour, skin milk powder, and casein.

Proteins are found in fractions of the egg but most of the proteins are in egg white and egg yolk. Egg white proteins have been used for various industrial applications, but egg yolk proteins (EYP) have not been paid much attention. Egg yolk is rich in proteins and lipoproteins. The lipid does not exist as free droplets but is associated with proteins in the form of lipoproteins [2]. Investigations of egg yolk proteins have identified the presence of phosvitin, livetin, lipovitellin, lipovitellenin, and some minor components. Fractionation of these components of yolk has indicated that they are all heterogeneous. However, a paucity of information still exists on their detailed structures.

II. CLASSES OF EGG YOLK PROTEINS

Lipoproteins are separated into high density lipoproteins (HDL), low density lipoproteins (LDL), and very low density lipoprotein (VLDL) by whether they float or settle in a solvent of chosen density (1.063 for HDL and LDL; 1.007 for LDL and VLDL) [3]. The details are mentioned below.

A. LIPOVITELLIN (HIGH DENSITY LIPOPROTEINS)

Egg yolk is a fluid emulsion with a continuous phase of protein (livetins) and a dispersed phase of lipoprotein particles, e.g., lipovitellins or HDL and lipovitellenins or LDL. Yolk proteins, thus, exist associating with lipids as lipoproteins. Egg yolk contains 15.3% protein and 31.2% lipids, and the components include, carbohydrates, minerals, and moisture, etc.

Lipovitellins and phosvitin are the proteolytic cleavage products of the precursor, vitellogenin, which is synthesized in the liver under the regulation of estrogenic hormones. Some of the known EYP are summarized in Table 1 [4]. Vitelline was the first yolk protein to be recognized [4].

Lipovitellin is one of the two lipoproteins contained in hen egg yolk and comprises about one sixth of the yolk solids. It is spherical in shape with a molecular weight of 4×10^5, composed of about 80% protein and 20% lipid [5]. The properties of lipovitellins have been, thus, extensively studied [5-7].

α– and β–lipovitellins are distinguished in electrophoresis and separated on hydroxyapatite or by ion exchange chromatography [6, 8]. α– and β–lipovitellins are composed of at least 8 polypeptides with similar molecular weights ranging from 35,000 to 140,000 daltons. Both lipovitellins contain mannose, galactose, glucosamine and sialic acid. The sialic acid content in α–lipovitellin exceeds that in β–lipovitellin by six times, though only slight differences were found in the content of neutral and amino sugars. The relative acidic nature of α–lipovitellin is also due to the predominance of sialic acid [9].

α– and β–lipovitellins differ in both lipid and phosphorus protein content, depending on methods of preparation. However, their amino acid and lipid compositions are quite similar. Vitelline, the apoprotein of lipovitellin, has 0.75% polysaccharide [10].

Table 1
Yolk Proteins

	% in Hen Eggs
Apovitellenin I (apoVLDL II)	0.7
Apovitellenin II	0.1
Apovitellenin III	0.1
Apovitellenin IV	0.6
Apovitellenin V	0.6
Apovitellenin VI	0.7
Biotin-binding protein	trace
Lipovitellin apoproteins	
α-Lipovitellin	2.0
β-Lipovitellin	1.0
Livetins	
α-Livetin (serum albumin)	0.2
β-Livetin ($α_2$ glycoprotein)	0.3
γ-Livetin (γ-globulin)	0.2
Phosvitin	1.0

Modified from Burley and Vadehra (1989) [4] by permission of John Wiley & Sons, New York.

Burley and Cook found 50% dissociation at pH 10.5 and 7.8 for α–lipovitellin dimers and β–lipovitellin dimers, respectively [7]. They also found that the dissociation process was reversible with both materials. Sulfhydryl and protein phosphorus groups are presumably not directly involved in the dissociation reactions. Employing sulfhydryl blocking agents and varying the pH, they found that there was no significant change in the monomer-dimer equilibria [11]. It was also established that dissociation of dimers decreased with temperature, which means that the strength of bonding between monomer units must increase with temperature, behavior characteristic of hydrophobic forces [5].

Lipovitellins appear to have a surface structure quite different from that of LDL. Burley and Kushner were able to remove 95% of the lipid phosphorus of LDL with phospholipase C, whereas they could liberate only 6% of the lipid phosphorus from α– and β–lipovitellins with the same enzyme [12].

The solubilities of lipovitellins are influenced by their lipid content. When lipovitellins are delipidated, they suffer a marked loss of solubility [5].

B. LIPOVITELLENIN (LOW DENSITY LIPOPROTEINS)

LDL in yolk plasma has a density of 0.98, and it should really be classified as VLDL [5]. LDL in yolk accounts for two-third of the total yolk solids, and it is composed of 89% lipid and 11% protein [13]. The lipid content of LDL is 70% triacylglycerol, 4% cholesterol, and

26% phospholipid [5].

LDL is prepared from fresh egg yolk by diluting it with an equal volume of distilled water. The resulting clots, after dissolving, are salted out with ammonium sulfate, applied on a DEAE-cellulose column and then eluted with 0.05 Tris-HCI buffer, pH 8.2 containing 10^{-3} M EDTA. The LDL solution is dialyzed with distilled water and then freeze dried [14].

Lipovitellenin does not exist as a separate entity from yolk. It is derived from the LDL of yolk plasma by ether extraction [15]. LDL treated with ether gives lipovitellenin; a fraction remains between the ether and water phases. LDL is heterogeneous. Fractionation of LDL produces LDL1, and LDL2, which differ slightly in chemical composition.

LDL1 and LDL2 have molecular weights of 10 and 3 million, respectively [16]. The two fractions are not represented evenly in yolk plasma: LDL1 is only about 20% of the total LDL but it contains approximately twice the amount of protein as LDL2 [16]. Ultracentrifugation revealed that LDL1 and LDL2 have five components each with different sedimentation rates [17]. The first 3 components have lipid contents of 50%, 67% and 44%, respectively. Components with lower sedimentation rates tend to aggregate and form components with higher sedimentation rates when lipid is gradually removed.

Lipovitellenin is soluble in an aqueous solution [18, 19]. Vitellenin, the apoprotein of lipovitellenin, is a glycoprotein with 2.45% carbohydrates containing 1.3% hexose, 0.67% hexosamine, and 0.38% sialic acid [19]. The amino acid composition of vitellenin has been analyzed and lysine appears to be the predominant *N*-terminal amino acid [20], whereas the C terminus is glutamic acid [21]. When digested with pronase, vitellenin gave two glycopeptides which separated on Sephadex G25 [19]. Glycopeptide A has a high content of sialic acid and is negatively charged. Glycopeptide B contains most of the carbohydrates of vitellenin, but it is devoid of sialic acid.

C. PHOSVITIN

Phosvitin is a non lipid phosphoglycoprotein of phosvitin (2.5% hexose, 1.0% hexosamine, and 2.0% sialic acid) [10]. It contains 12% nitrogen and 10% phosphorus (representing 80% of the protein phosphorus in yolk).

Phosvitin, the most highly phosphorylated protein, represents 7% of egg yolk powder. This phosphoprotein is the second most plentiful macromolecular constituent of avian egg yolk granules and is one of the main proteins of hen eggs. An unusual, very acidic, proteinlike material was discovered in chicken egg yolk more than 100 years ago. Subsequently known as "vitellinic acid," it was found to contain high levels of phosphate esterified to serine.

Mecham and Olcott described a procedure for a purified protein that they termed "phosvitin" [22]. Although these early phosvitin preparations appeared homogenous in the ultracentrifuge [23], free solution electrophoresis indicated that one, two, or possibly three components were present. By use of stepwise salt elution from an anion exchange column, Connelly and Taborsky, separated phosvitin into two subfractions (a major "0.30" and minor "0.35 phosvitin") [24]. Yolk proteins, e.g., lipovitellin and phosvitin are proteolytically derived within the growing oocyte.

The purified preparations of phosvitin, the phosphoprotein of hen egg yolk, contain 5.5-6.5% carbohydrate. Analysis showed 6 residues of hexose, 5 of glucosamine, and 2 of sialic acid per molecular weight of 40,000. In order to determine the mode of attachment of the polysaccharide moiety, phosvitin has been digested with pronase [25, 26].

The amino acid composition of phosvitin has been analyzed. There are very little or no sulfur-containing amino acids. However, serine constitutes 31% of the total amino acid residues and is present in the form of phosphoserine. Consecutive residues of phosphoserine up to 8 units have been observed in phosvitin [27]. These consecutive phosphoserine residues are believed to play an important role in the stability and iron-binding properties of phosvitin [27]. The N-terminal amino acid of phosvitin is alanine [28], but the sequencing of amino acids is still unknown.

Pure phosvitin is soluble in water [29]. Phosvitin is homogeneous under centrifugation but shows up as two components when analyzed by electrophoresis or chromatography [24, 30, 31]. Using gel filtration on Sephadex G–200, it was found that the two components of phosvitin, α–phosvitin and β–phosvitin, have molecular weights of 160,000 and 190,000 [32]. Phosvitin polypeptides react readily to form aggregates in an aqueous solution. The two phosvitins also differ considerably in phosphorus and carbohydrate contents.

Egg is well known to decrease the absorption of iron. The low availability of iron may be due to its tight binding to the phosvitin molecule through the mechanism by which the latter inhibits iron absorption. Two moles of organic phosphorous in phosvitin can bind one mole of iron. The 20% egg yolk protein diet would contain enough phosvitin to bind all iron in the diet. Phosvitin is a protein well suited for the binding of metals including iron. In its metallic complexed form, phosvitin shows a dynamic reactive behavior with significant potential for biological exploitation. In transit from the liver to the oocyte, it certainly functions as an iron carrier. In the egg itself, it is a major component of a highly structured particle and its role in forming and maintaining this organized yolk element is undoubtedly assisted, if not regulated, by its metal complex. Phosvitin, by virtue of its poly anionic character, binds readily with metal ions of more than one charge to form aggregates of increased molecular weight [25].

Iron binding provides a contrast in a number of ways. Spectroscopic properties of the ferric phosvitin complex have been interpreted in terms of a polynuclear iron cluster bound tetrahedrally to phosphate ligands such that pairs of phosphates are brought into effective proximity [33].

Phosvitin is unordered at "physiological pH" but acquires a β–type conformation when the electrostatic repulsion between the many negatively charged phosphate groups is eased by a suppression of their ionization as the pH [33] or dielectric constant is lowered. Yolk iron is concentrated in this fraction. Other yolk components carrying cationic charges could also interact with the phosphoprotein and induce a transition to an ordered structure.

Whole egg contains 1.97 mg of iron/100 g, and 95% of the iron is bound to phosvitin which indicates that the iron is probably bound in a tetrahedral coordination structure with oxygen in the phosphoserine residues [27, 34]. Phosvitin contains 135 residues of phosphoserine. Phosvitin can bind one iron atom for every two phosphate groups [33, 35]. Therefore, the maximum Fe/P ratio of 0.5 is possible. Phosvitin in egg yolk is not saturated with iron, and it should be able to bind additional iron provided in a diet.

Phosvitin is relatively resistant to attack by pepsin and trypsin [36]. It is not easily denatured by cooking. A strong chelating agent (0.03M EDTA) currently being used in some food preparations, is capable of releasing 50% of the bound iron from phosvitin [27].

D. LIVETIN

Plimmer (1908) termed "livetin," an anagram of vitelline for a protein fraction in ether-treated egg yolk that remains in solution after the precipitation of the rest of the yolk solids with water [37].

Livetin is a water-soluble, non lipid, globular glycoprotein in yolk plasma. It has 15% nitrogen and 1.8% sulfur, and it represents 4-11% of total solids in yolk [36]. Using moving–boundary electrophoresis, Shepard and Hottle resolved livetin into 3 fractions [38]. Martin and his co-workers later discovered that the α and β fractions had 15 bands when analyzed with disc gel electrophoresis, but the γ–fraction showed only 1 band [39]. William identified α–livetin as serum albumin, β–livetin as α–glycoprotein, and γ–livetin as γ–globulin corresponding to IgG in mammals [40]. Since γ–livetin is in egg yolk, it is now called IgY.

The mixed livetins can be isolated from diluted yolk by gel filtration chromatography on columns of agarose that permit the lipoprotein to pass through (Biogel A5m) [4]. Methods for isolating livetin often involve extraction with organic solvents or high speed centrifugation [6].

As the α–, β– and γ–livetin correspond to serum proteins which have been more thoroughly studied than those of yolk, useful information can be derived from the properties of these proteins.

α– livetin is sensitive to high salt concentrations, α– and β– livetins are sensitive to organic solvents, and γ–livetin is especially sensitive to high temperatures [4]. Livetin contains IgG (mol. wt. 180,000) and is believed to be a promising source of these antibodies [41]. A simple improved method of isolation and details of its application IgY has recently been reported, and the details are described in Chapter 11.

III. CHARACTERISTIC STRUCTURES OF EGG YOLK PROTEINS

The HDL, LDL or VLDL, and phosvitin are present in various structures. Spheres, profiles, myelin figures (MF), and granules are characteristic of egg yolk proteins.

Spheres are a minor component of egg yolk and they contain droplets which have been identified as lipoproteins [42] and oil drops [43]. Most spheres have a diameter of between 20 μm and 40 μm [44].

Profiles are round, about 25 nm in diameter, and are uniformly distributed in yolk as individual particles. Profiles are considered hydrated LDL [44].

Myelin figures have dimensions of 45-110 nm by 56-130 nm [45] and are identified as VLDL (MF) [46]. Pure MF have been successfully isolated from granules [47, 48], and it is postulated that MF micelles are "particles consisting of an inner neutral lipid core stabilized with one or more lamellae composed of phospholipid, cholesterol, and protein."

The apoproteins of VLDL were found to be similar in polypeptide composition to that of LDL in yolk plasma (supernatant fraction of yolk after centrifugation) [14, 49, 50].

Chemical analysis of granules indicated the presence of 60% protein, 34% lipid, and 6% ash [44]. Lipids in granules are represented by 37% phospholipids of which 82% is phosphatidylcholine and 15% is phosphatidylethanolamine [44].

Burley and Kushner (1963) removed 95% of the lipid phosphorus of LDL using phospholipase C and concluded that the phospholipid was probably located at the surface of the LDL micelles. Because phospholipid is virtually insoluble in triglyceride [12], The core of the micelles is likely to be composed of triglycerides [51]. Also, proteins are presumed to be on the surface of the micelles, because 80% of the protein of LDL is accessible to proteolytic enzymes [52, 53]. Cholesterol is also probably found on the surface of the micelles, because cholesterol is normally associated with phospholipids in biological materials [54].

The three sedimenting fractions (upon centrifugation of granules) consist of many strands to which microparticles are attached. The strands are phosvitin complexes whereas the microparticles are lipovitellin. The microparticles are electrophoretically negative indicating the abundant presence of phospholipids. Phosvitin and lipovitellin have a strong affinity for each other and the bond between them is believed to be a salt linkage elaborated by electrostatic reaction [54]. Knowledge of the characteristic structures of proteins and lipoproteins in yolk is essential to their successful isolation and characterization.

IV. EMULSIFYING PROPERTIES OF EGG YOLK PROTEINS

Egg yolk is recognized as an excellent emulsifying agent, and it is used extensively in food preparations; e.g., mayonnaise and salad dressings [55]. The emulsifying power of egg yolk is attributable to the presence of lipoproteins. However, efficient emulsifiers are complexes in which proteins and lipoproteins are important components [55].

Mizutani and Nakamura have investigated the emulsifying power of egg yolk lipoprotein and found that the protein component of LDL is probably the main contributor to the emulsifying activity, because of the decreasing emulsifying activity of LDL with increased proteinase treatment [56-58]. They also noted that, at low LDL or protein concentrations, only a small amount of LDL or protein was necessary to significantly increase emulsifying activity; whereas at high LDL or protein concentrations, additional LDL or protein has little or no effect on emulsifying activity.

LDL is a major component of hen egg yolk and contains 86-89% lipids [59]. Its molecular structure is considered to be a large spherical particle with a core of triglyceride and a surface layer of both phospholipid and proteins. Emulsifying properties of LDL are due to the characteristic structure of its lipid protein complexes.

Apoprotein can be obtained from LDL through delipidation with organic solvents, but the protein obtained is not soluble in water, although various kinds of detergents and salts were used for preparation of apoprotein in a soluble state from the LDL of human plasma [60, 61]. Removal of residual detergents and salts from the proteins has remained a difficult problem, because it affects emulsifying properties remarkably.

In emulsions prepared with egg yolk LDL, gradual removal of protein with proteinases does not substantially decrease emulsion stability. Lipids, including phospholipids and cholesterol, are also important components in an emulsion. Mizutani and Nakamura have demonstrated that increased emulsifying activity and emulsion stability are associated with increased amounts of lipids bound to LDL [58]. Emulsifying activity and emulsion stability of egg yolk are described in detail in Chapter 8.

V. GELLING ABILITY OF EGG YOLK PROTEINS

The irreversible increase in the viscosity of egg yolk as a result of freezing and thawing, treatment with heat, or some chemicals, is due to gelation, a kind of irreversible denaturation of protein. Egg yolk is often stored in the frozen state before being used. The freeze-thaw gelation is generally regarded as undersirable, because it renders the yolk difficult to handle in automated processing [19]. Freeze-thaw gelation occurs when yolk is stored below -6°C [62]. The rate and extent of gelation is influenced by the rate of freezing, the temperature and duration of storage in the frozen state, and the rate of thawing.

The mechanism of freeze-thaw gelation is still obscure. It appears that freezing causes the formation of ice crystals. At -6°C, about 81% of the initial water in yolk is frozen, producing ice crystals [63]. The proteins become less hydrated when water is removed from the solution. The breakdown of the water surrounding the protein molecules could promote rearrangement by causing aggregation of EYP to result in a gel [64]. There is also evidence that gelation may result from the breaking of lipid-protein bonds in LDL [65]. Delipidation of egg yolk plasma using phospholipase C, n-heptane, or 1-butanol has been demonstrated to cause gel formation [66]. It is hypothesized that, when the lipid is removed, the areas on the protein molecules normally bound to the lipid interact with one another resulting in a gel [65].

Heat-induced gelation is often an important functional property of proteins, especially, of egg white protein. Heating causes unfolding of protein molecules. The functional groups are, thus, exposed and attracted to one another resulting in a gel [67]. The heat-induced gelation of egg proteins is also mentioned in Chapter 8.

VI. HEAT STABILITY OF EGG YOLK PROTEINS

EYP are heat labile except for phosvitin. Mecham and Olcott subjected solutions of phosvitin to pH 4-8 at 100°C but they observed no precipitate or any evidence of molecular changes [36]. Phosvitin is extremely heat stable at temperatures below 110°C. The heated samples of phosvitin in 0.02M imidazole-HCl (pH 9), at 10°C intervals from 80 to 140°C showed no occurrence of gelation up to 140°C with little or no liberation of phosphorus up to 110°C. In gel electrophoresis, the heated samples showed two distinct α– and β–phosvitin components up to 110°C [32].

Low temperature spray drying does not affect the functional properties of EYP. High temperature (>55°C), direct, gas-heated spray drying should not be applied for EYP, since it induces the formation of radicals in the powder via oxidative reaction involving either oxygen and/or nitrogen oxides to cause a remarkable denaturation of EYP [68]. EYP are treated in order to dissociate the aggregates and to dry the resulting solution to a white powder.

VII. COMMERCIAL PRODUCTS-EGG YOLK PROTEIN

A. PREPARATION OF EYP-80

Hen eggs were broken and egg yolk was collected. A commercial preparation of egg yolk protein (EYP-80) was carefully separated and dried after delipidation under mild conditions in our facilities (Figure 1). EYP-80 contains around 80% proteins besides some lipids and carbohydrates (Table 2).

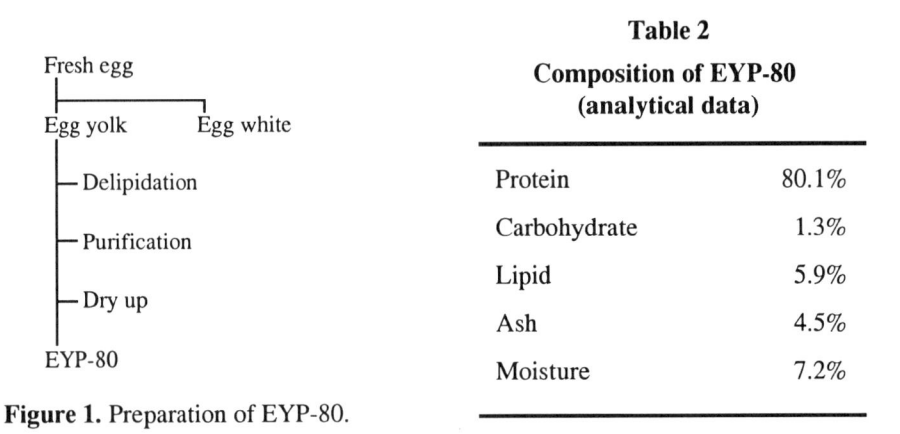

Figure 1. Preparation of EYP-80.

Table 2
Composition of EYP-80 (analytical data)

Protein	80.1%
Carbohydrate	1.3%
Lipid	5.9%
Ash	4.5%
Moisture	7.2%

B. AMINO ACID PATTERN OF EYP-80

The quality of protein is generally determined by comparing the essential amino acids. EYP-80 contains all nine essential amino acids. Their amounts in mg of the preparation are more than recommended by FAO/WHO/UNU for 2-5 year-old children [69]. This is the most demanding pattern of any age group other than infants (Figure 2). The amino acid score of EYP-80 was found to be 100.

C. MINERALS AND VITAMINS OF EYP-80

EYP-80 is an admirable source of indispensable minerals (free or bound in organic

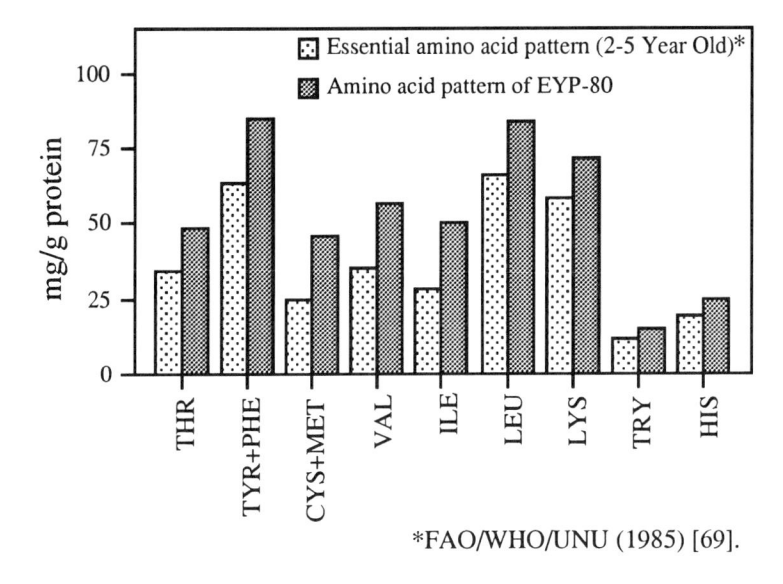

Figure 2. Amino acid pattern of EYP-80.

molecules). EYP-80 contains calcium (770 mg/100g). Also, it contains less sodium and has a relatively high content of potassium, which is an important feature from a nutritional point of view. EYP-80 has a significant amount of iron, magnesium and selenium also (Table 3). Selenium is an important physiological antioxidant. Low levels of selenium affect glutathione peroxides activity which may lead to cancer in human beings [22, 70, 71]. EYP-80 furnishes large quantities of readily available organic phosphorus. Phosvitin contains 10% phosphorus, representing about 7% of the EYP-80 and about 80% of the EYP-80-bound phosphorus.

EYP-80 contains vitamin A, D, E, B12, choline, folic acid, vitamin B3, biotin, inositol, and vitamin B6 but not vitamin C. Some of the vitamins measured in EYP-80 are mentioned in Table 3.

Table 3
Minerals and Vitamins of EYP-80

Minerals (mg/100g)		Vitamins (mg/100g)	
Na	190	Total Tocopherol	0.60
K	370	Vitamin B6	0.76
Ca	770	Pantothenate	11.5
Mg	81	Folate	5.8
P	930	Choline	210
Fe	38	Biotin	0.20
Zn	18		
I	0.22		
Se	0.15		

D. ANTIOXIDANT ACTIVITY OF EYP-80

To measure the antioxidant effect of EYP-80, 15 ml of DHA-enriched fish oil (containing 27% DHA) dissolved in 150 ml of diethyl ether was mixed with various ratios of EYP-80. DHA-enriched EYP-80 powder was obtained by drying the mixture in N_2 gas. After the oxidative treatment, the oil fraction from the EYP-80 powder was extracted. The peroxide value (PV) [72] and carbonyl value (CV) [73] were measured. EYP-80 was found to be antioxidative based on the concentration used against DHA-enriched fish oil (Figure 3). The extent of oxidation was evaluated by the increase in PV and CV values with EYP-80 as compared to a control at different temperatures (Figure 4) and under different illuminance confirmed the antioxidant activity of EYP-80.

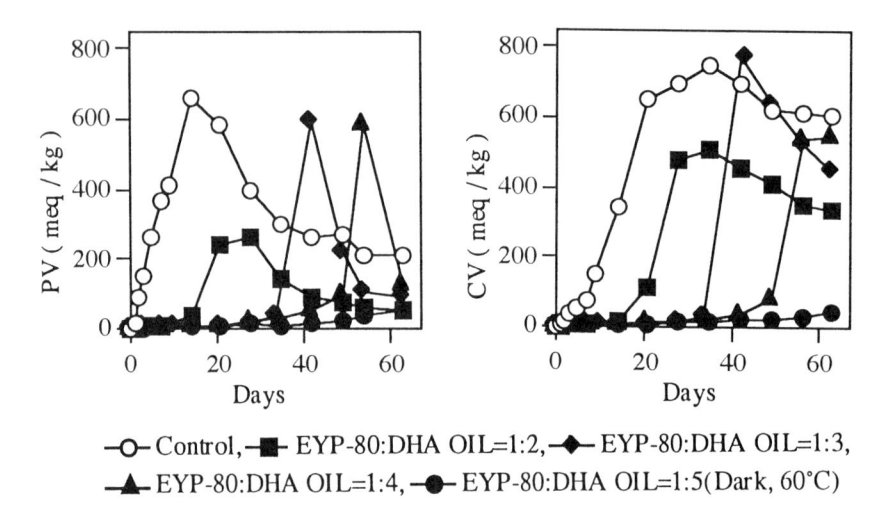

Figure 3. Effect of EYP-80 concentration on oxidation of DHA oil.

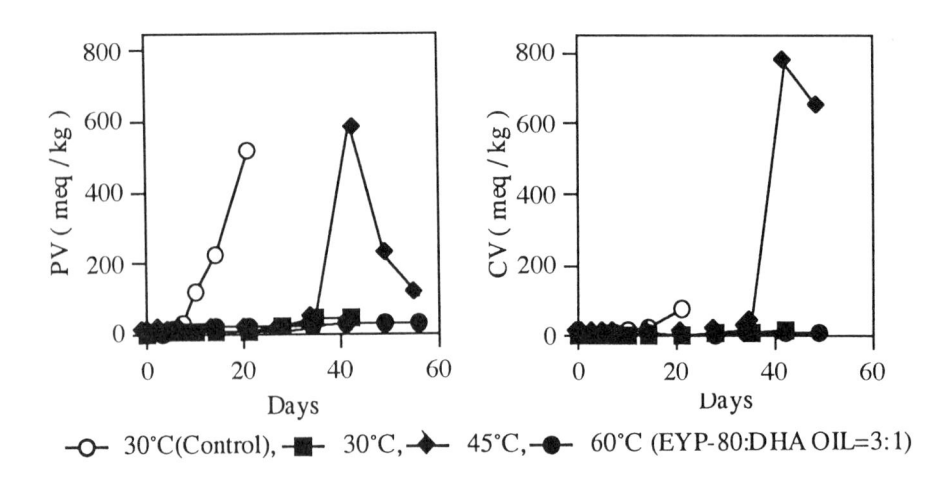

Figure 4. Effect of EYP-80 on oxidation of DHA oil at different temperatures.

E. CHOLESTEROL CONTENT OF EYP-80

A growing concern for health has driven an interest in protein sources that are low in fat and cholesterol. Although vegetable proteins do not have cholesterol, EYP-80 is nutritionally superior to those. Some proteins from vegetable sources are known to contain some anti-institutional factors such as trypsin inhibitors, phytates, and estrogens, etc. [74]. EYP-80 was purified to contain less than 0.1% (1/30) cholesterol as compared to 2.6% in egg yolk powder (Figure 5).

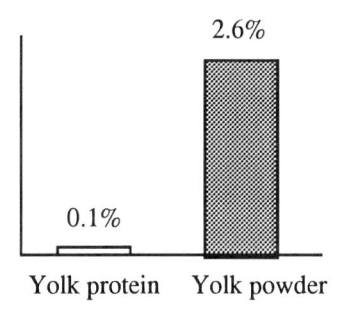

Figure 5. Cholesterol content of egg yolk powder and EYP-80.

VIII. CONCLUDING REMARKS

Egg yolk contains fascinating functional and immunologically functional proteins of which we have investigated the technology for manufacturing a dried preparation to be effectively applicable for food and other industrial fields. The functional egg yolk proteins, prepared in a powder state as EYP-80, which is characterized by a potent antioxidative activity with almost no content of cholesterol, would give a new choice for the designer of foods looking for novel functional ingredients.

Before the advent of infant formulas, egg yolk was recommended by most physicians primarily as a source of proteins and iron for infants. The efficacy of such a recommendation may be in doubt in view of the heat stability and iron-binding properties of phosvitin. However, our understanding of the egg yolk proteins is limited. The egg yolk is a nutrient reservoir for the growing chick embryo which apparently has no problem making use of the iron stored in phosvitin.

The chicken egg is a relatively inexpensive, renewable source. Increasingly, uses are being made of its various components. The yolk is rich in proteins and lipoproteins. To make maximal use of these yolk components, we must further elucidate their structural-functional relationships. EYP-80 should be superior to proteins of various cereals and milk in maintaining the amino acid balance in humans.

REFERENCES

1. **Romanoff, A. L. and Romanoff, A. J.,** Food value, in *The Avian Egg,* Romanoff, A. L. and Romanoff, A. J., Eds., John Wiley & Sons, New York, 1949, p 375.

2. **Parkinson, T. L.**, The chemical composition of eggs, *J. Sci. Food Agri.*, **17**, 101, 1966.
3. **Lindgren, F. T. and Nichols, A. V.**, Structure and function of human serum lipoproteins, in *The Plasma proteins,* Vol. 2, Putnam, F. W., Ed., Academic Press, New York, 1960.
4. **Burley, R. W and Vadehra, D. V.**, *The Avian Egg. Chemistry and Biology,* Burley, R. W. and Vadehra, D. V., Eds., John Wiley & Sons, New York, 1989.
5. **Cook, W. H. and Martin, W. G.**, Egg Lipoproteins, in *Structural and Functional Aspects of Lipoproteins in Living Systems,* Tria, E. and Scanu, A. M., Eds., Academic Press, New York, 1969.
6. **Bernardi, G. and Cook, W. H.**, An electrophoretic and ultracentrifugal study on the proteins of the high density fraction of egg yolk, *Biochim. Biophys. Acta*, **44**, 86, 1960.
7. **Burley, K. W. and Cook, W. H.**, The dissociation of α- and β-lipovitellins in aqueous solution. Part I. Effect of pH, temperature, and other factors, *Can. J. Biochem. Physiol.*, **40**, 362, 1962.
8. **Radomski, M. W. and Cook, W. H.**, Fractionation and dissociation of the avian lipovitellins and their interaction with phosvitin, *Can. J. Biochem.*, **42**, 395, 1964.
9. **Kurisaki, J., Yamauchi, K., Isshiki, H., and Ogiwara, S.**, Difference between α- and β-lipovitellin from hen egg yolk, *Agric. Biol. Chem.*, **45**, 699, 1981.
10. **Ito, Y. and Fujii, T.**, Chemical compositions of the egg yolk lipoproteins, *J. Biochem. (Tokyo)*, **52**, 221, 1962.
11. **Burley, R. W. and Cook, W. H.**, The dissociation of α- and β-lipovitellins in aqueous solution. Part II. Influence of protein phosphate groups, sulhydryl groups, and related factors, *Can. J. Biochem. Physiol.*, **40**, 373, 1962.
12. **Burley, R. W. and Kushner, D. J.**, The action of *Clostridium perfringens* phosphatidase on the lipovitellin's and other egg constituents, *Can. J. Biochem. Physiol.*, **41**, 409, 1963.
13. **Cook, W. H.**, Proteins of hen's egg yolk, *Nature*, **190**, 1173, 1961.
14. **Raju, K. S. and Mahadevan, S.**, Isolation of hen's egg yolk very low density lipoproteins by DEAE-cellulose chromatography, *Anal. Biochem.*, **61**, 538, 1974.
15. **Martin, W. G., Tattrice, N. H., and Cook, W. H.**, Lipid extraction and distribution studies of egg yolk lipoproteins, *Can. J. Biochem. Physiol.*, **41**, 657, 1963.
16. **Martin, W. G., Augustyniak, J., and Cook, W. H.**, Fractionation and characterization of the low density lipoproteins of hen's egg yolk, *Biochim. Biophys. Acta*, **84**, 714, 1964.
17. **Augustyniak, J., Martin, W. G., and Cook, W. H.**, Characterization of lipovitellenin components, and their relation to low-density lipoprotein structure, *Biochim. Biophys. Acta*, **84**, 721, 1964.
18. **Turner, K. J. and Cook, W. H.**, Molecular weight and physical properties of a lipoprotein from the floating fraction of egg yolk, *Can. J. Biochem. Physiol.*, **36**, 937, 1958.
19. **Vadehra, D. V. and Nath, K. R.**, Eggs as a source of protein, *Crit. Rev. Food Technol.*, **4**, 193, 1973.
20. **Smith, D. B. and Turner, K. J.**, *N*-terminal amino acids of the proteins of the floating fraction of egg yolk, *Can. J. Biochem. Physiol.*, **36**, 951, 1958.
21. **Neelin, J. M. and Cook, W. H.**, Terminal amino acids of egg yolk lipoproteins, *Can. J. Biochem. Physiol.*, **39**, 1075, 1961.
22. **Mecham, D. K. and Olcott, H. S.**, Phosvitin, the principal phosphoprotein of egg yolk, *J. Am. Chem. Soc.*, **71**, 3670, 1949.
23. **Wallace, R. A. and Morgan, J. P.**, Isolation of phosvitin: retention of small molecular weight species and staining characteristics on electrophoretic gels, *Anal. Biochem.*, **157**, 256, 1986.
24. **Connelly, C. and Taborsky, G.**, Chromatographic fractionation of phosvitin, *J. Biol. Chem.*, **236**, 1364, 1961.

25. **Clark, R. C.,** Amino acid sequence of a cyanogen bromide cleavage peptide from hens egg phosvitin, *Biochim. Biophys. Acta,* **310,** 174, 1973.
26. **Clark, R. C.,** The primary structure of avian phosvitins, *Int. J. Biochem.,* **17,** 983, 1985.
27. **Albright, K. J., Gordon, D. T., and Cotterill, O. J.,** Release of iron from phosvitin by heat and food additives, *J. Food Sci.,* **49,** 78, 1984.
28. **Freeney, R. E.,** Egg proteins, in *Symposium of Foods: Proteins and Their Reactions,* Schultz, H. W. and Anglemier, A. F., Eds., AVI Publishing Co., Inc., Westport, Connecticut, 1963.
29. **Cook, W. H.,** Macromolecular Components of Egg Yolk, in *Egg Quality: A Study of the Hen's Egg,* Carter, T. C., Ed., Oliver & Boyd, Edingburgh, 1968.
30. **Sugano, H.,** Studies on egg yolk proteins. II. Electrophoretic studies on phosvitin, lipovitellin, and lipovitellenin, *J. Biochem.,* **44,** 205, 1957.
31. **Abe, Y., Itoh, T., and Adachi, S.,** Fractionation and characterization of hen's egg yolk phosvitin, *J. Food Sci.,* **47,** 1903, 1982.
32. **Ito, T., Abe, Y., and Adachi, S.,** Comparative studies on the α- and β-phosvitin from hen's egg yolk, *J. Food Sci.,* **49,** 1755, 1983.
33. **Taborsky, G.,** Iron binding by phosvitin and its conformational consequences, *J. Biol. Chem.,* **255,** 2976, 1980.
34. **Webb, J., Multani, J. S., Saltman, P., Beach, N. A., and Gray, H. B.,** Spectroscopic and magnetic studies of iron phosvitins, *Biochemistry,* **12,** 1797, 1973.
35. **Taborsky, G. and Mok, C.-C.,** Phosvitin: Homogeneity and molecular weight, *J. Biol. Chem.,* **242,** 1495, 1967.
36. **Williams, T.,** Serum proteins and the livetins of hen's egg yolk, *Biochem. J.,* **83,** 346, 1962.
37. **Plimmer, R. H. A.,** The proteins of egg yolk, *Trans. Chem. Soc. London,* **93,** 1500, 1908.
38. **Shepard, C. C. and Hottle, G. A.,** Studies of the composition of the livetin fraction of the yolk of hen eggs with the use of electrophoretic analysis, *J. Biol. Chem.,* **179,** 349, 1949.
39. **Martin, W. G., Vandegaer, J. E., and Cook, W. H.,** Fractionation of livetin and the molecular weights of α- and β-components, *Can. J. Biochem. Physiol.,* **35,** 241, 1957.
40. **Williams, J.,** Serum proteins and the livetins of hen's egg yolk, *Biochem. J.,* **83,** 346, 1962.
41. **Jensenius, J. C., Andersen, I., Hau, J., Crone, M., and Koch, C.,** Eggs: Conveniently packaged antibodies, methods for purification of yolk IgG, *J. Immol. Methods,* **46,** 63, 1981.
42. **Bellairs, R.,** The structure of the yolk of the hen's egg as studied by electron microscopy. I. The yolk of the unincubated egg, *J. Biophys. Biochem. Cytol.,* **11,** 207, 1961.
43. **Grodzinski, Z.,** The yolk spheres of the hen's egg as osmometers, *Biol. Rev.,* **26,** 253, 1951.
44. **Powrie, W. D.,** Chemistry of Eggs and Egg Products, in *Egg Science & Technology (2nd Ed.),* Stadehman, W. J. and Cotterill, O. J. Eds., AVI Publishing Co., Inc., Westport, Conneticut, 1977.
45. **Chang, C. M., Powrie, W. D., and Fennema, O.,** Microstructure of egg yolk, *J. Food Sci.,* **42,** 1193, 1977.
46. **Kocal, J. T., Nakai, S., and Powrie, W. D.,** Preparation of apolipoprotein of very low density lipoprotein from egg yolk granules, *J. Food Sci.,* **45,** 1761, 1980.
47. **Garland, T. D. and Powrie, W. D.,** Isolation of myelin figures and low-density lipoproteins from egg yolk granules, *J. Food Sci.,* **43,** 592, 1978.
48. **Garland, T. D. and Powrie, W. D.,** Chemical characterization of egg yolk myelin figures and low-density lipoproteins isolated from egg yolk granules, *J. Food Sci.,* **43,** 1210, 1978.

49. **Kocal, J. T., Nakai, S., and Powrie, W. D.,** Chemical and physical properties of apoprotein of very low density lipoprotein from egg yolk granules, *J. Food Sci.*, **45**, 1756, 1980.
50. **Raju, K. S. and Mahadevan, S.,** Protein components in the very low density lipoproteins of hen's egg yolks: Identification of highly aggregating (gelling) and less aggregating (non gelling) proteins, *Biochim. Biophys. Acta*, **446**, 387, 1976.
51. **Schneider, H. and Tattrice, N. H.,** Mutual solubility of the lipid components of egg yolk low-density lipoprotein, *Can. J. Biochem.*, **46**, 979, 1968.
52. **Steer, D. C., Martin, W. G., and Cook, W. H.,** Structural investigation of the low-density lipoprotein of hen's egg yolk using proteolysis, *Biochemistry*, **7**, 3309, 1968.
53. **Margolis, S. and Langdon, R. G.,** Studies on human serum β_1-lipoprotein, *J. Biol. Chem.*, **241**, 485, 1966.
54. **Margolis, S.,** Structure of very low and low density lipoproteins, in *Structural and Functional Aspects of Lipoproteins in Living Systems,* Tria, E. and Scanu, A. M., Eds., Academic Press, London and New York, 1969.
55. **Baldwin, R. E.,** Functional Properties in Foods, in *Egg Science and Technology (2nd Ed.),* Stadleman, W. J. and Cotterill, O. J., Eds., AVI Publishing Co., Inc., Westport, Conneticut, 1977.
56. **Mizutani, R. and Nakamura, R.,** Emulsifying properties of egg yolk low density lipoprotein (LDL): Compared with bovine serum albumin and egg lecithin, *Lebensm. - Wiss. u. -Technol.*, **17**, 213, 1984.
57. **Mizutani, R. and Nakamura, R.,** The contribution of polypeptide moiety on the emulsifying properties of egg yolk low density lipoprotein (LDL), *Lebensm. -Wiss. u. -Technol.*, **18**, 60, 1985.
58. **Mizutani, R. and Nakamura, R.,** Physical state of the dispersed phases of emulsions prepared with egg yolk low density lipoprotein and bovine serum albumin, *J. Food Sci.*, **50**, 1621, 1985.
59. **Mizutani, R. and Nakamura, R.,** Emulsifying properties of a complex between apoprotein from hen's egg yolk low density lipoprotein and egg yolk lecithin, *Agric. Biol. Chem.*, **51**, 1115, 1987.
60. **Steele, J. C. H. and Reynolds, J. A.,** Characterization of the apolipoprotein B polypeptide of human plasma low density lipoprotein in detergent and denaturant solutions, *J. Biol. Chem.*, **254**, 1633, 1979.
61. **Helnius, A. and Simon, K.,** Removal of lipids from human plasma low-density lipoprotein by detergents, *Biochemistry*, **10**, 2542, 1971.
62. **Morris, E. R. and Greene, F. G.,** Utilization of the iron of egg yolk for hemoglobin formation by the growing rat, *J. Nutr.*, **102**, 901, 1972.
63. **Powrie, W. D. and Nakai, S.,** Characteristics of Edible Fluids of Animal Origin: Eggs, in *Food Chemistry (2nd Ed.),* Fennema, O. R., Ed., Marcel Dekker, Inc., New York, 1985.
64. **Powrie, W. D.,** Gelation of Egg Yolk upon Freezing and Thawing, in *Low Temperature Biology of Foodstuffs,* Hawthorn, J. and Rolfe, E. J. Eds., Pergamon Press, Oxford, 1968.
65. **Mahadevan, S., Satyanarayana, T. and Kumar, S. A.,** Physicochemical studies on the gelation of hen's egg yolk: Separation of gelling protein components from yolk plasma, *J. Agric. Food Chem.*, **17**, 767, 1969.
66. **Kumar, S. A. and Mahadevan, S.,** Physicochemical studies on the gelation of hen's egg yolk: Delipidation of yolk plasma by treatment with phospholipase C, and extraction with solvents, *J. Agric. Food Chem.*, **18**, 666, 1970.
67. **Nakamura, R., Fukano, T., and Taniguchi, M.,** Heat-induced gelation of hen's egg yolk low-density lipoprotein (LDL) dispersions, *J. Food Sci.*, **47**, 1449, 1982.
68. **Morgan, J. N. and Armstrong, D. J.,** Quantification of cholesterol oxidation products in

egg yolk powder spray dried with direct heating, *J. Food Sci.*, **59**, 43, 1992.
69. **FAO/WHO/UNU.**, Energy and protein requirements, *Reports of the Joint FAO/WHO/UNU Expert Consultations (FAO, WHO and the United Nations University)*, Geneva, Switzerland, 1985.
70. **Yoshida, M., Yasumoto, K., Iwami, K., and Tashiro, H.**, Distribution of selenium in bovine milk and selenium deficiency in rats fed casein-based diets, monitored by lipid peroxidase level and glutathione peroxidase activity, *Agric. Biol. Chem.*, **45**, 1681, 1981.
71. **Willett, W. C., Morris, J.S., Pressel, S., Taylor, J.O., Polk, B.F., Stampfer, M.J., Rosner, B., Schneider, K., and Hames, C.G.**, Prediagnostic serum selenium and risk of cancer, *Lancet*, **73**, 171, 1982.
72. **Yamamoto, Y. and Omori, M.**, Antioxidative activity of egg yolk lipoproteins, *Biosci. Biotech. Biochem.*, **58**, 1711, 1994.
73. **Chiba, T., Takazawa, M., and Fujimoto, K.**, A simple method for estimating carbonyl content in peroxide containing oils, *J. Am. Oil Chem. Soc.*, **66**, 1588, 1989.
74. **Ananthraman, K. and Finot, P.**, Nutritional aspects of food proteins in relation to technology, *Food Reviews International*, **9**, 629, 1993.

Chapter 6

EGG YOLK LIPIDS

L.R. Juneja

TABLE OF CONTENTS

I. Introduction
II. Classes of Egg Yolk Lipids
    A. Neutral Lipids
    B. Phospholipids
        1. Structure of Phospholipids
        2. Major Components of Egg Yolk Phospholipids
            a. Phosphatidylcholine
            b. Phosphatidylethanolamine
            c. Lysophosphatidylcholine
            d. Sphingomyelins
    C. Other Components of Egg Yolk Lipids
        1. Cholesterol
        2. Gangliosides
    D. Occurrence of Phospholipids
III. Yolk Lipid Extraction And Fractionation
    A. Egg Yolk Extraction
        1. Fresh Egg Yolk Extraction
        2. Egg Yolk Powder Extraction
    B. Column Chromatographic Fractionation of Egg Yolk Lipids
        1. Silica Gel Column Chromatography
        2. Ion Exchange Cellulose Chromatography
IV. Enzymatic Conversion of Phospholipids
V. Analysis of Egg Yolk Lipids
VI. Dietary Effects on Egg Yolk Lipids
    A. Dietary Fatty Acids
    B. Dietary Cholesterol
VII. Applications of Egg Yolk Lipids
    A. Egg Yolk Lipids as a Source of Choline
    B. Egg Yolk Lipids as a Source of Arachidonic Acid and Docosahexaenoic Acid
    C. Egg Yolk Lipids as a Source of Cholesterol
    D. Egg Yolk Lipids as an Antioxidant
    E. Egg Yolk Lipids as a Constituent of Liposome (Pharmaceuticals)
    F. Egg Yolk Lipids in Cosmetics
VIII. Conclusion

References

I. INTRODUCTION

The name "lecithin" is derived from the Greek name "Lekithos" corresponding to egg yolk, from which the lipid was the first isolated. The mature egg yolk of the domestic hen egg processes a remarkably constant lipid and lipoprotein composition depending upon the feed composition. The great uniformity in the composition of the egg yolk lipids (EYL) serves to maintain its chemical and physical properties in various industrial applications.

Avian eggs differ from most other animal eggs in the high proportion of yolk lipids. Dry yolk contains 60% lipids (Figure 1). There is little doubt that the main function of yolk lipids is to provide metabolic energy for the embryo. In fresh eggs, the lipids are combined noncovalently with protein, largely in particles of lipoprotein, and to a very small extent with carbohydrates. Metabolism and mechanism of deposition of lipoproteins has served a valid area of scientific inquiry in its own right.

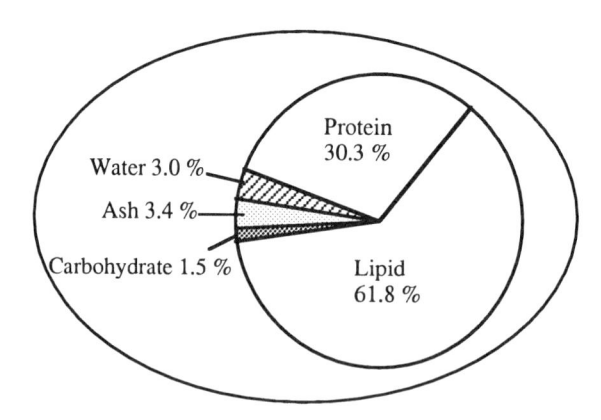

Figure 1. Composition of egg yolk powder.

Phospholipids (PLs) are natural biosurfactants as well as important material for building cells and tissue. Also, PLs have many applications in the cosmetic, food, and pharmaceutical industries [1-5]. In living microorganisms, PLs act in flow and transport processes through biological membranes. Thus, they are mediators between the surrounding medium and the cytoplasm as well as between cytoplasm and the cell compartments.

EYL have been popularly used in the manufacture of pharmaceutical emulsions. Research into the structure and function of biological membranes and the development and application of liposomes have led to increasing use of PLs in recent years [6]. Highly purified PLs are necessary for the solution of pressing biochemical and medical problems, for example, the creation of highly specific methods to diagnose some infectious diseases. More and more PLs of high purity with a certain fatty acid composition and definite polar head groups are being demanded.

II. CLASSES OF EGG YOLK LIPIDS

Triacylglycerols (TGs) and PLs constitute the main components of EYL. The main feature of lipids which affect their solubility in organic solvents are the nonpolar hydrocarbons chains of the fatty acid moieties and some functional groups such as phosphate or sugar residues. Thus, neutral lipids are nonpolar and highly soluble in hydrocarbon solvents such as hexane or benzene. In more polar solvents, however, they are quite insoluble. The contrary will be true of PLs. PLs are quite polar as a result of the choline or ethanolamine molecule esterified to phosphoric acid. EYL contains about 65% neutral lipids (EYL-N), 31% PLs, and 4% cholesterol (Figure 2).

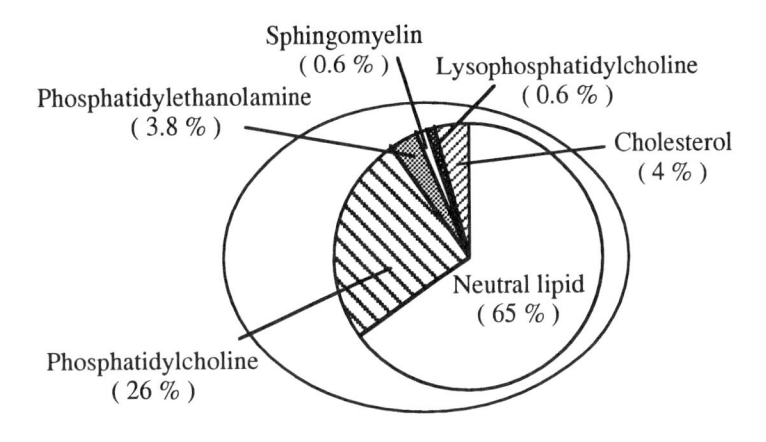

Figure 2. Composition of egg yolk lipids.

A. NEUTRAL LIPIDS

An average egg provides about 6g of lipids which are contained exclusively in the yolk. About two-thirds of the total dry yolk mass is fats. Thus, yolk can be regarded as a potentially important oil source.

Early workers have used silicic acid chromatography to separate and quantitate neutral lipid classes in egg yolk lipid extracts. The first extensive analysis was reported by Privette and his co-workers [7]. Besides TGs, the other constituents of neutral lipids include diacylglycerols, monoacylglycerols, fatty acids, carotenoids, sterols, etc. Position 1 of the TG molecule is mostly occupied by palmitic acid (saturated acid), position 2 by oleic and linoleic acids (unsaturated acids), and position 3 by oleic, palmitic, and stearic acids (unsaturated and saturated acids), respectively [8].

B. PHOSPHOLIPIDS
1. Structure of phospholipids

The lipids which contain glycerol as a backbone, TGs and PLs, are named glycerolipids. PLs are lipids which contain phosphate and have a glycerol-phosphate backbone. Baer and Kates (1950) by chemical synthesis have shown that lecithin is based on L-α glycerophosphate like other naturally occurring glycerophospholipids [9]. The nomenclature of PLs has undergone numerous modifications in the last three decades. The latest recommendations are those by the IUPAC-IUB Commission on Biochemical Nomenclature, and the Stereospecific numbering system is now used for all the PLs. It suggests a stereo specific numbering system, in which for glycerol "the carbon atom that appears on top in the Fischer projection that shows a vertical carbon chain with the secondary hydroxyl group to the left is designated as Carbon 1 (C-1)."

To indicate such numbering, the prefix *sn-* is used (Figure 3). Carbon 2 (C-2) is a chiral center (asymmetric carbon). The hydroxyl groups on C-1 and C-2 are usually acylated with fatty acids (nonpolar tails). In most PLs, the fatty acid substituent at C-1 is saturated and at C-2 is unsaturated (Figure 4).

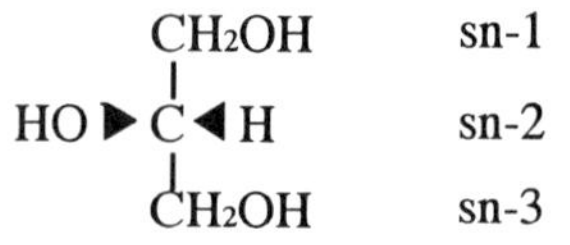

$$\begin{array}{ll} CH_2OH & sn\text{-}1 \\ HO \blacktriangleright C \blacktriangleleft H & sn\text{-}2 \\ CH_2OH & sn\text{-}3 \end{array}$$

Figure 3. Stereospecific numbering of glycerol derivatives.

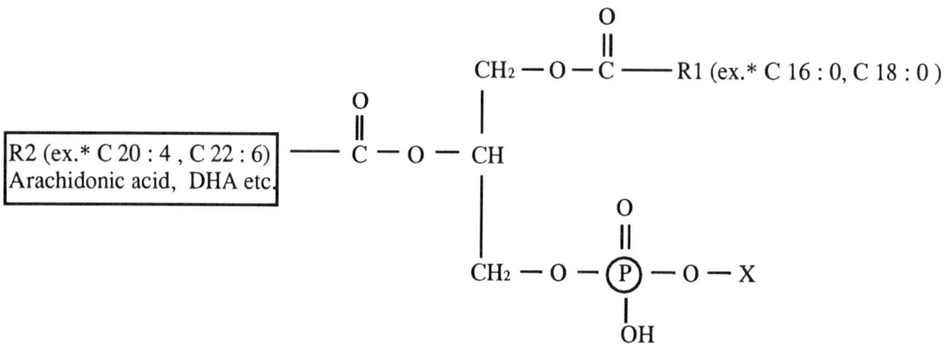

Figure 4. Basic structure of egg yolk phospholipid.
*ex. = example

The PLs are classified according to the polar head group (X) on the phosphate group. If X is a hydrogen, the compound is called 3-*sn*-phosphatidic acid (PA). If X is choline, the PL is called PC (the recommended name: 1,2-diacyl-*sn*-glycero-3-phosphocholine or 3-*sn*-phoshatidycholine (PC)). Other PLs, i.e., phosphatidylethanolamine (PE), phosphatidylglycerol (PG), phosphatidylserine (PS), and phosphatidylinositol (PI), are also named according to their X group (Figure 5).

In nature, PLs often exist as complex mixtures with the same X group (e.g., choline) but with various kinds of different acyl substituents. For example, human red blood cells have 21 different molecular species of PC that differ in fatty acid substituents at either C-1 or C-2, or both. Although PLs are divided into six major classes (Figure 5), the molecular species reach more than 100. The advantage of such complexity is not readily apparent but fatty acid composition is important for membrane-associated active transport.

The PLs are amphiphilic molecules, because they have both polar and nonpolar groups. The polar head group prefers an aqueous environment (hydrophilic), whereas the nonpolar acyl groups gives hydrophobicity. Therefore, most PLs spontaneously form a bilayer when suspended in an aqueous environment, and the bilayer is thought to be responsible for the major structural organization of most membranes in which proteins are embedded.

X = −H Phosphatidic acid (PA)

X = −CH₂CH₂N⁺(CH₃)₃ Phosphatidylcholine (PC)

X = −CH₂CH₂NH₂ Phosphatidylethanolamine (PE)

X = −CH₂CH(OH)−CH₂OH Phosphatidylglycerol (PG)

X = −CH₂CH(NH₂)−COOH Phosphatidylserine (PS)

X = [inositol ring] Phosphatidylinositol (PI)

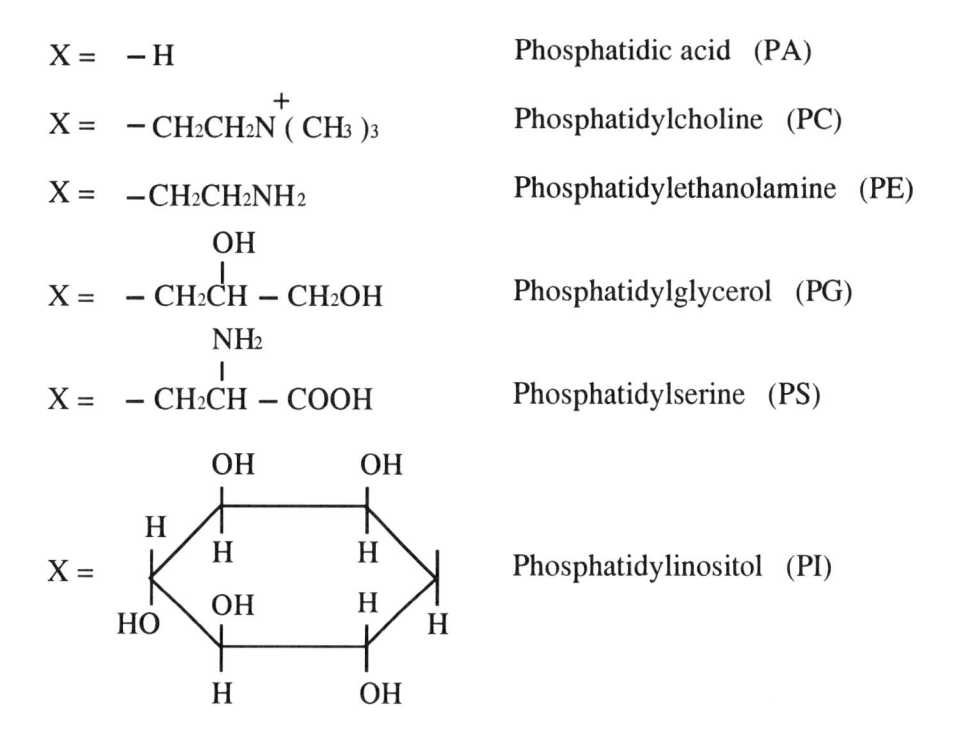

Figure 5. Major polar head moities of phospholipids.

2. Major components of egg yolk phospholipids

a. Phosphatidylcholine (PC)

PC constitutes around 80% of the total PLs and is the largest class of PLs of egg yolk (Table 1). The chemical composition and physicochemical properties of egg yolk PC have been extensively investigated. It has been shown that the saturated fatty acids are located primarily in the *sn*-1 and the unsaturated fatty acids in the *sn*-2 position of PLs [10-13]. Fatty acid composition of two common PCs showed that C16 and C18:2 are the major fatty acids in egg and soybean PC, respectively [14]. However, polyunsaturated fatty acids, e.g., arachidonic acid (AA), and docosahexaenoic acid (DHA), are found in egg yolk PC and other yolk PLs but not in soy PC [14] (Table 2).

b. Phosphatidylethanolamine (PE)

PE is the second most abundant PL (11.9%) in egg yolk and frequently accompanies PC in commercial preparations of egg yolk lecithin (Table 1). Thudichum (1884) separated a nitrogen- and phosphorus-containing lipid fraction from brain tissue. He distinguished this fraction from lecithin by its relative insolubility in warm ethanol and named it "Kephalin" from which ethanolamine was obtained by hydrolysis [15]. Palmitic, stearic, oleic, and AA are the major fatty acids of PE, but linoleic and DHA are also present in significant amounts [16].

c. Lysophosphatidylcholine (LPC)

Lysophosphatidylcholine (LPC) is a minor constituent (1.9%) of egg yolk of PLs in which one of the acyl substituents (usually from C-2) is missing. The lysophospholipids are named by simply adding the prefix lyso- to the name of the PLs.

d. Sphingomyelins (SM)

Sphingomyelins (SM) are present in EYL and most of the animal cell membranes. SM (N-acyl sphingosine 1-phosphocholine) is a minor component of egg yolk PLs. Sphingolipids are based on a backbone of sphingosine, a long chain alcohol that contains a hydrophobic hydrocarbon tail, an amine group where a fatty acyl group may attach, and a C-1 hydroxyl group where a polar head group attaches. The addition of a fatty acyl group to the amino group of sphingosine produces a sphingolipid called ceramide. The addition of a phosphate group to the C-1 hydroxyl group of ceramide and attachment of a choline to this phosphate produces SM. The composition of egg yolk SM and ceramides have been determined for the long chain base and fatty acid composition [17]. The egg yolk SM is made up largely of medium chain length fatty acids, with the palmitoylsphingosines accounting for about 85% of the total SM [18, 19].

Table 1

Composition of Egg Yolk Lecithin

Phospholipid	Concentration %
Phosphatidylcholine	80.8
Phosphatidylethanolamine	11.7
Lysophosphatidylcholine	1.9
Sphingomyelin	1.9
Neutral Lipids + Others	3.7

Reproduced from Juneja et al. (1994) [52] by permission of CAB International, Oxon.

Table 2

Fatty Acid Composition of Phosphatidylcholine Extracted from Egg and Soybean*

Fatty Acid Composition	Egg PC	Soy PC*
16:0 Palmitic acid	32	12
16:1 Palmitoleic acid	1.5	<0.2
18:0 Stearic acid	16	2.3
18:1 Oleic acid	26	10
18:2 Linoleic acid	13	68
18:3 Linolenic acid	<0.3	5
20:4 Arachidonic acid	4.8	<0.1
22:6 Docosahexaenoic acid	4	<0.1

*Weiner et. al. (1989) [14].

C. OTHER COMPONENTS OF EGG YOLK LIPIDS
1. Cholesterol
The major sterol in EYL is cholesterol, but traces of cholesterol oxides [20] and plant sterols [21] are also found. Essentially all cholesterol is found in the free form. The estimates of total cholesterol range from 11-15 mg/g yolk [20-22] which is 3-5% of the total PLs (Figure 2), and the cholesterol content is only slightly changed by cholesterol in the feed [22].

2. Gangliosides
The molecular species of the gangliosides from egg yolk have been resolved by TLC [23]. The major species are found to be N-acetylneuraminosyllactosylceramide and di-N-acetylneuraminosyllactosylceramide. Galactosylceramide is the only neutral glycosphingolipid found in chicken egg yolk [23].

D. OCCURRENCE OF PHOSPHOLIPIDS
PLs are a distinct class of biomolecules present in all types of cells where they constitute the major components of membrane structures. Usually PC is the main constituent of natural PLs whereas the content of other components, such as PE, PI, PG, PS, and PA, vary with the source.

Oil in vegetables, such as soybean, corn, cottonseed, linseed, peanut, rapeseed, safflower, and sunflower, have been used as lecithin sources [23]. Soybean oil is one of the main sources of commercial lecithin, and, besides PLs, it contains TGs, carbohydrates, and sterols, etc. [24, 25]. PC, the most important PL, is the most abundant found in egg yolk. The PL composition of microorganisms is quite different from those of plants and animals [26]. The main PL in *E. coli* is PE. The compositions of lecithins from some different sources are summarized in Table 3. The PL content of mammalian cells varies from organ to organ and from species to species [26].

Table 3

Compositions (%) of Lecithins (Fat Free) from Various Sources

PLs	Egg yolk*	Rapeseed**	Soybean**	Bovine brain**	*Escherichia coli***
PC	84	37	27	27	—
PE	12	29	22	36	80
PI	—	14	19	2	—
PS	—	—	1	18	Trace
PG	—	—	—	—	15
PA	—	—	15	2	—
SM	2	—	—	15	—
GLs[a]	—	20	12	—	—
CL[b]	—	—	—	—	5
LPLs[c]	3	—	4	—	—

[a] Glycolipids; [b] cardiolipin; [c] lysophosholipids.

*Juneja (unpublished results).

**Ansell et. al. (1982) [26].

III. YOLK LIPID EXTRACTION AND FRACTIONATION

The routine method used for the extraction of PLs from fresh egg yolk is a modification of the method of Pangborn [27]. If one starts with liquid egg yolk, the preferred solvent is acetone which first removes the water and then the TGs of yolk, leaving acetone-insoluble PLs and proteins. This mixture is then split by alcohols to extract the phosphorous-containing lipids [28]. A somewhat exceptional process for obtaining egg lecithin is by extraction with dimethyl ether under moderate pressure in a closed vessel [29]. In principle, the extraction processes for biological materials to obtain lipid mixtures similar to the lecithin composition mentioned in the literature are innumerable [30-50]. In the past few years, a new technology has been discussed in the literature [51]. This is the treatment of lipid mixtures with supercritical gases but it is not economical because of the high equipment cost. The drawback of available techniques to prepare pure PC is the difficulty of removal of LPC and SM from PC. A method has been established for large scale preparation of chromatographically homogeneous, high purity PLs using an ion-exchange column [52]

A. EGG YOLK EXTRACTION
1. Fresh egg yolk extraction

Fresh egg yolk (500 kg) was blended thoroughly with acetone (1 *l*) at 25°C, the mixture was allowed to stand for 1h, and then filtered. The acetone extract contained most of the neutral lipids and pigment. The solids were washed thrice with 1,500*l* of acetone and then suspended in 1,250*l* ethanol, mixed and allowed to stand for 1 h and the mixture was filtered. The extraction was repeated with 1,250*l* of ethanol and the combined ethanol extracts were concentrated to dryness under reduced pressure. Fresh egg yolk (500 kg) treated with acetone and alcohol yielded 37.5 kg of the lecithin containing 95% PLs based on the total PLs present in the egg yolk. The egg yolk lecithin obtained was found to contain a high PC content (80.8%) (Table 1). The average yield of extracted PLs in the above extraction procedure was 7.2% based on the weight of fresh egg yolk.

2. Egg yolk powder extraction

Egg yolk powder (1.0 ton) was extracted with 14-fold ethanol in two extractions at 40°C. The moisture content of ethanol was carefully controlled and the ethanol recovered was recycled. The EYL-30 obtained was dried and further purified by acetone to obtain EYL-65 and EYL-95. A simplified procedure of EYL preparation is summarized in Figure 6. The analytical values of the EYL products EYL-N, EYL-30, EYL-65, and EYL-95 obtained, which include PL content, acid value, iodine value, and peroxide value, are mentioned in Table 4.

Table 4

Analytical Values of Egg Yolk Lipids

Parameter	EYL Products			
	EYL-N	EYL-30	EYL-65	EYL-95
Phospholipids	<1%	31.30%	68.00%	96.80%
Acid Value	1.4	6.3	15	19.3
Iodine Value	82.0	75.2	68.7%	62.1
Peroxide Value	2	<0.1	<0.1	0.1
Loss on Drying	1.80%	0.50%	2.5%	1.30%
Heavy Metals	<20ppm	<20ppm	<20ppm	<10ppm
Arsenic	<2ppm	<2ppm	<2ppm	<2ppm

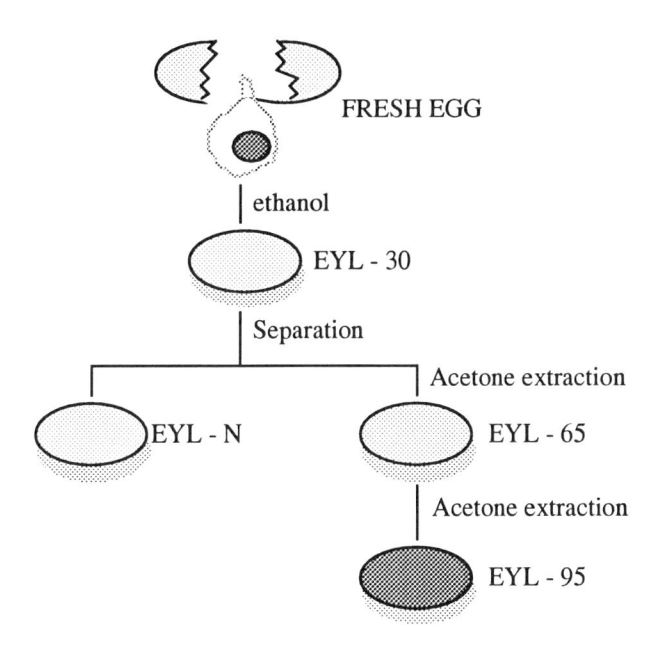

Figure 6. Preparation of egg yolk lipids.

B. COLUMN CHROMATOGRAPHIC FRACTIONATION OF EGG YOLK LIPIDS
1. Silica gel column chromatography

EYL-95 (>95% PLs containing >80% PC) was applied to a silica gel column (Si 15-30 mm; Soken Kagaku, Japan), and PLs were separated with methanol/water (98:2), as an eluent, at a flow rate of 200 ml/min (Figure 7). The separation of EYL-95 (200g) on the silica gel yielded 22.5g of PE (>99% purity) and 81.1g of PC (>99% purity).

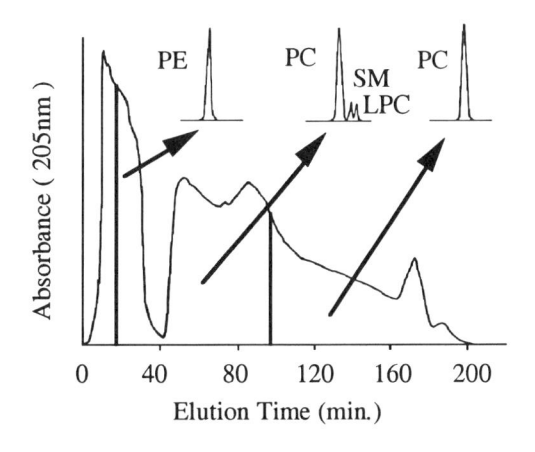

Figure 7. Separation of egg yolk lipids (200g EYL-95).
Column: Silica gel column (15-30μm, 100 × 500 mm).
Eluent : MeOH / H2O = 98 / 2
Flow rate : 200 ml / min
UV detection : λ 205 nm
Reproduced from Juneja et al. (1994) [52] by permission of CAB International, Oxon.

2. Ion exchange cellulose chromatography

The fraction containing choline PLs (2g), i.e., PC mixed with LPC and SM obtained from the silica gel chromatography was further fractionated on the ion exchange cellulose column (500mm x 20mm i.d.) using dichloromethane/methanol (9:1) stepwise elution at a flow rate of 10 ml/min (Figure 8). The purity of PC, PE, and LPC >98% (yield 70-80%) and the purity of SM at 92.7% (yield 76%) was obtained. The fatty acid/phosphorous molar ratio of the purified PLs was 2.0.

The overall recovery of egg yolk PC having purity of 99.2% was 81.8%. The purity of PE was 99.3% with a recovery of 79.8%. The purity of LPC and SM was 98.9% and 92.7%, respectively.

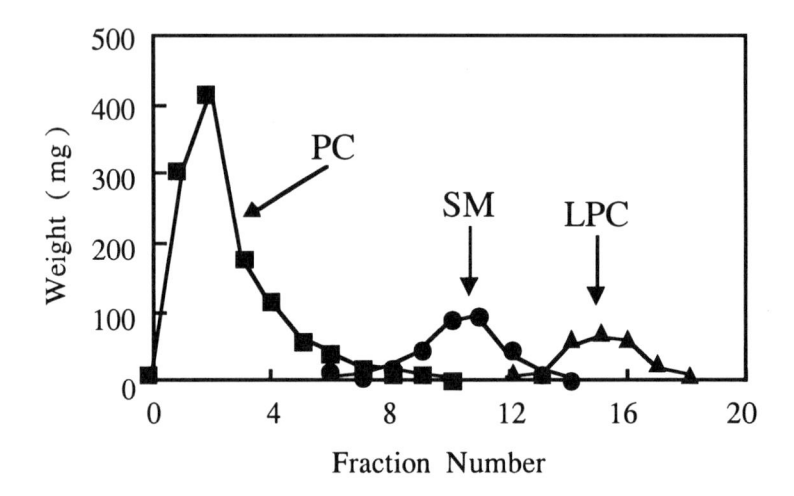

Figure 8. Separation of phosphatidylcholine from lysophosphatidylcholine and sphingomyelin.
Column: Ion exchange cellulose
Eluent: Dichloromethane/methanol (stepwise elution)
Flow rate: 10 ml/min

IV. ENZYMATIC CONVERSION OF PHOSPHOLIPIDS

Various PLs which are not found in egg yolk can be obtained by enzymatic transphosphatidylation of PC in the presence of respective acceptors (Figure 9).

The transphosphatidylation reaction has been performed for PG, PE, or PS synthesis from PC in the biphasic reaction mixture. PC was added to ethyl acetate and the mixture was sonicated (Sonicator, Ohtake works, Japan; continuous pulse for 3 min) under ice-cold conditions in a sonication vessel. Phospholipase D from *Streptomyces* in an aqueous buffer (desired pH) containing glycerol, ethanolamine, or serine was introduced into the reactor to which PC and ethyl acetate were added. The reaction was carried out at 30°C with continuous stirring [4].

As an example, the conversion of PC to PS by phospholipase D preparation was studied in the presence of L-serine or D-serine. At a concentration of 3.4 M serine which was the saturated solution of serine in a buffer of pH 5.6 at 30°C, phospholipase D yielded 99.8% and 99.5% of PS using 17.8 mM PC with L-serine and D-serine, respectively (Table 5). Almost no byproduct (PA) as formed, giving a selectivity of approximately 100%. Similar yields have been obtained with PG and PE formation from PC [53].

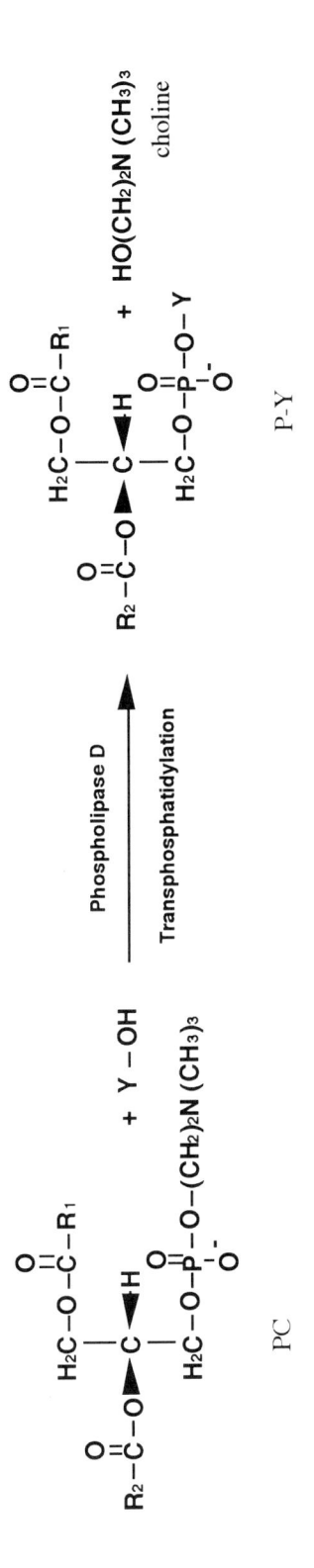

$$\begin{array}{c} O \\ \| \\ R_2-C-O \end{array} \begin{array}{c} H_2C-O-C-R_1 \\ \| \\ O \\ \blacktriangle C \blacktriangledown H \\ H_2C-O-P-O-(CH_2)_2N(CH_3)_3 \\ | \\ O^- \end{array} \quad + \quad Y-OH \quad \xrightarrow[\text{Transphosphatidylation}]{\text{Phospholipase D}} \quad \begin{array}{c} O \\ \| \\ H_2C-O-C-R_1 \\ | \\ R_2-C-O \blacktriangle C \blacktriangledown H \\ \| \\ O \\ H_2C-O-P-O-Y \\ | \\ O^- \end{array} \quad + \quad HO(CH_2)_2N(CH_3)_3$$

PC
choline

P-Y

1. Y-OH: Glycerol, ethanolamine, or serine, etc.

2. Selectivity = $\dfrac{[\text{Product}]}{[\text{Product}] + [\text{Byproduct}]} \times 100$

3. [Product] = PS, PE, PG, or PI
   [Byproduct] = PA

Figure 9. Enzymatic conversion of polar head group of phospholipids. Reproduced from Juneja et al. (1994) [52] by permission of CAB International, Oxon.

Table 5
Enzymatic Conversion of Polar Group of Phospholipids*

Conversion	Substrate[a] conc. (mM)	Source of PLD	Inhibition by high conc. of acceptor[b] (%)	Conversion (%)	Selectivity (%)
PC→PG	17.8	Cabbage	+	100	100
PC→PE	17.8	Streptomyces	+	100	100
PC→PS	17.8	Streptomyces	−[c]	100	100
PC→PI	17.8	Streptomyces	−[c]	0	0

[a] Substrate used was PC.

[b] Acceptors used were glyceol or ethanolamine or L-serine or D-serine or myo inositol.

[c] Limited solubility of the acceptor in aqueous phase.

*Juneja (1989) [53]

V. ANALYSIS OF EGG YOLK LIPIDS

PLs have been analyzed mainly by thin-layer chromatography [53], thin-layer chromatography/flame-ionization detection [54], high-performance liquid chromatography [55, 56], and gas chromatography [57]. Recently, phospholipases have been used to determine the structure of unidentified PLs [58].

VI. DIETARY EFFECTS ON EGG YOLK LIPIDS

The total dietary fat of hens has little effect on the overall content of yolk lipids. However, the fatty acid composition of EYL reflects that of the dietary fat to a large extent.

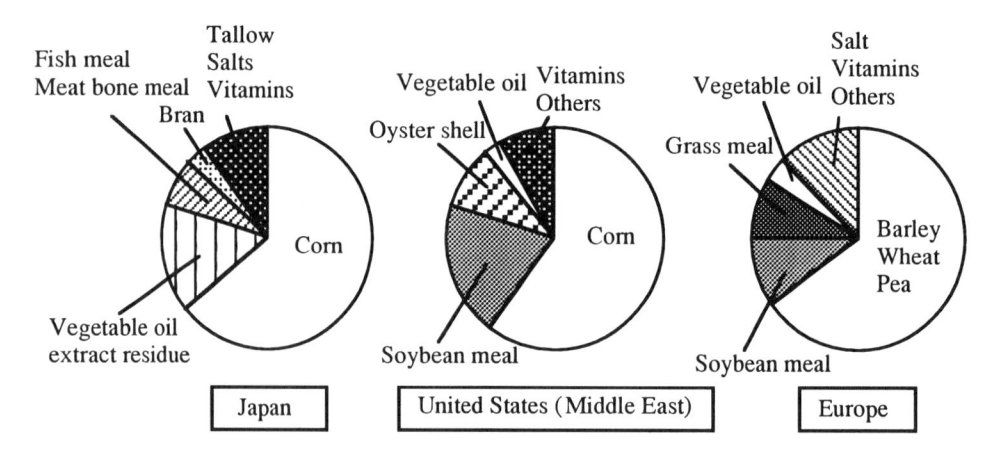

Figure 10. Hen feed used in several parts of the world.

A. DIETARY FATTY ACIDS

Omega eggs (DHA-enriched eggs) are becoming popular as a food for health. The ω3 fatty acid content is raised to conform with good health. The *sn*-2 acyl groups of egg yolk PLs reflects the change, when the chickens are maintained on diets containing unsaturated fats. Omega (DHA) eggs produced by hens ingesting a specially formulated feed contains a high level of ω3 fatty acids. Eggs produced by hens on ordinary feed contain 0.3%-0.5% DHA, whereas eggs produced on specially formulated feed contain around 5-10 times higher levels of DHA [59].

It is known that the fatty acid composition of egg PLs can be altered through dietary manipulation. Jiang and Sim have reported that the longer chain ω3. PUFA were preferentially deposited into the PL fraction, particularly into PE fraction of the yolk [60]. Full fat oil seeds, e.g., flax seed and high oleic and linoleic acid seeds, were selected as the dietary source to change the fatty acid composition. Hens fed on different diets in Europe, the United States and Japan (Figure 10) have shown the change in their fatty acid composition (Table 6).

Table 6

Arachidonic Acid and Docosahexaenoic Acid
Contents of Eggs from Various Markets of The World

Production of eggs in	Yolk Lipids (EYL-30)	
	Arachidonic Acid (%)	Docosahexaenoic Acid (%)
USA	1.50-1.84	0.54-0.70
Japan	1.60-1.90	0.90-1.07
Europe	1.40-1.85	0.99-1.20

B. DIETARY CHOLESTEROL

Dietary cholesterol affects the yolk cholesterol to some extent but not significantly [61]. Safflower oil feeding suppresses egg yolk cholesterol levels more than hydrogenated coconut oil feeding to hens. Naber (1970) reported that diethyl-aminoethyldiphenylvalerate decreases yolk cholesterol but does not replace it [62].

The relative resistance of the egg composition to alterations in diet apparently reflects the nutritional and structural requirements for the development of the embryo.

VII. APPLICATIONS OF EGG YOLK LIPIDS

EYL are natural surfactants which have many applications in the cosmetic, pharmaceutical, food, and other industries. Recently, the demand for modified lecithin (mixture of PLs) having a higher content of PC has been increasing. Pure PLs or their lyso derivatives alter the hydrophilic-lipophilic balance and enhance oil/water type emulsification compared to commercial lecithin [63]. The general PLs application areas include confectioneries (chocolate, etc.), snack foods, fast foods, noodles, bakery products, margarines, dairy products, ice cream, yogurt, milk drinks, meat and poultry processing. These are the fields in which PLs were applied as emulsifying, wetting, and dispersing agents, releasing agents, sealants, and lubricants etc. [63]. Recently, utilization of PLs has been increasing in nonfood products, such as soaps, detergents, dyes, fertilizers, leather, paints, papers, and textiles. Some of the major applications of EYL are mentioned below.

A. EGG YOLK LIPIDS AS A SOURCE OF CHOLINE

EYL extracted from egg yolk having around 25% PLs have a high content of PC (80% of total PLs), which is more than three fold higher than natural soy lecithin (Figure 11). More than 90% of egg yolk PLs are choline derivatives, in this connection, PC, LPC, and SM, etc. PC is a major PL in brain and other tissues.

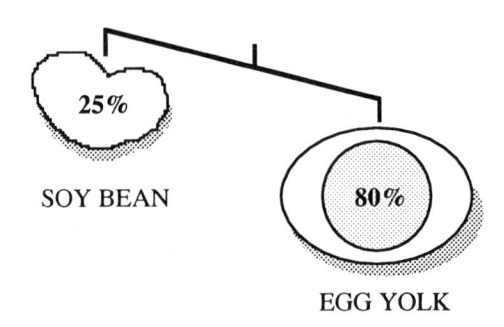

Figure 11. PC content of egg yolk and soybean lecithin.

Mammals derive a major fraction of choline from dietary sources. Less than 1% of choline normally present in the diet occurs as the free base. Most of the remainder is in the form of PC which is the major component of lecithin [63]. Consumption of a single meal containing lecithin, the major source of choline present naturally in the diet, increases the concentration of choline and acetylcholine in rat brain and adrenal gland [64]. PC consumption increases plasma and brain choline levels and accelerates neuronal acetylcholine synthesis. Hence, the concentration of acetylcholine in the tissue may normally be under direct, short term nutritional control. Consumption for 3 to 11 days of a diet supplemented with choline sequentially increases the concentration of serum choline, brain choline, and brain acetylcholine in rats [64]. Such precursor-induced changes in brain and adrenomedullary acetylcholine concentration are probably associated with parallel alterations in neurotransmitter release [65]. That similar increases in brain acetylcholine occur after ingesting choline is suggested by choline's utility in treating tardive dyskinesia, a brain disease thought to be associated with inadequate acetylcholine release [66]. EYL have high content of PC which should be a necessity for infant brain function. Recently, because of health consciousness and a dieting boom, many people in the United States and the Western world are reducing fat in their diets. Unfortunately, it leads to low intake of PLs and choline which is not a healthy trend. Choline is an essential nutrient for everyone [67]. Even a few weeks on a choline-deficient diet can cause abnormal liver in adults. PC protects against fibrosis in alcohol-fed baboons [68]. An inadequate choline intake might lower folic acid levels in women, endangering a current or future pregnancy, and choline deficiency in animals has been associated with infertility, growth retardation, and bone abnormalities [65, 69-75]. Choline enhanced learning in young rats when fed to the mothers before birth and to young rats after being born [65].

Choline is critical in cell communications, the process by which cells work together. Defects in cell communication lead to cancer, etc. [75]. Disruption of the cholinergic system is a major factor in producing the cognitive impairment that occurs in patients with Alzheimer's disease which is the most common form of dementia [76]. Administration of acetylcholine precursors has been one treatment approach [77]. Therefore, treatment with egg yolk lipids, a good source of choline (precursors of acetylcholine) would alleviate the symptoms of Alzheimer's disease.

Choline is routinely added to commercial infant formulations as an essential nutrient. The functions of PC as a source of choline are mentioned in Figure 12.

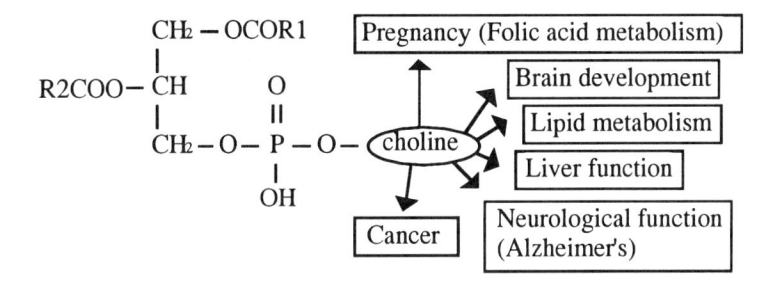

Figure 12. Functions of phosphatidylcholine as a source of choline.

Table 7

Fatty Acid Content of Egg Yolk Lipids

Fatty Acid (%)		EYL-30	EYL-65	EYL-95
Myristic Acid	C14:0	0.38	0.25	0.14
Palmitic Acid	C16:0	26.27	26.07	26.58
Palmitoleic Acid	C16:1	2.8	1.17	0.87
Margaric Acid	C17:0	0.32	0.33	0.31
Heptadeceanoic Acid	C17:1	0.36	0.22	0.15
Stearic Acid	C18:0	9.17	12.87	16.79
Oleic acid	C18:1	41.5	32.37	26.11
Linoleic Acid	C18:2 (ω-6)	13.24	16.42	15.61
Linolenic acid	C18:3 (ω-3)	0.3	0.56	0.12
Arachidonic Acid	C20:4 (ω-6)	1.73	3.94	6.03
Docosatetraenoic Acid	C22:4 (ω-6)	0.53	0.53	1.43
Docosahexaenoic Acid	C22:6 (ω-3)	0.51	3.16	4.54
Unknown		2.89	2.11	2.32

B. EGG YOLK LIPIDS AS A SOURCE OF ARACHIDONIC ACID AND DOCOSAHEXAENOIC ACID

EYL have unique fatty acids like AA and DHA (Table 7) which are not found in lecithin of soy and other plant origins (Table 2). AA and DHA content increases based on the contents of PLs (Table 7) which shows that AA and DHA are attached to PLs in EYL. Omega-3 fatty acids (DHA, etc.) are now regarded as essential in the diet for brain function and visual acuity in humans. Dietary ω-3 fatty acid deficiency led to visual loss in infant rhesus monkeys [78]. The ω-3 fatty acid content of infant formulas is an essential component and has a specific function in the photoreceptor membranes of the retina. AA is involved in maintaining membrane physicochemical properties and is a substrate for biosynthesis of a complex series of potent cell regulators referred to as eicosanoids [79, 80]. DHA and AA attached to PLs are important. Preterm infants can be at risk for poor growth due to low AA [80]. The ratio of DHA and AA is very important for the growth of infants. AA and DHA are found in high proportions in cell membranes, particularly those of the central nervous system and are present in human milk but not in the vegetable oils used in infant formulas. An average of fatty acids in human milk should cover the lipid requirements of infants [79-81]. Because of the lack of desaturating activity, infants cannot synthesize DHA and AA [82] (Figure 13). Therefore, there is a good reason to add DHA and AA content in infant formulas and design an improved fatty acid balance to meet the C20 and C22 polyunsaturated fatty acid (PUFA) requirements of infants (Figure 14).

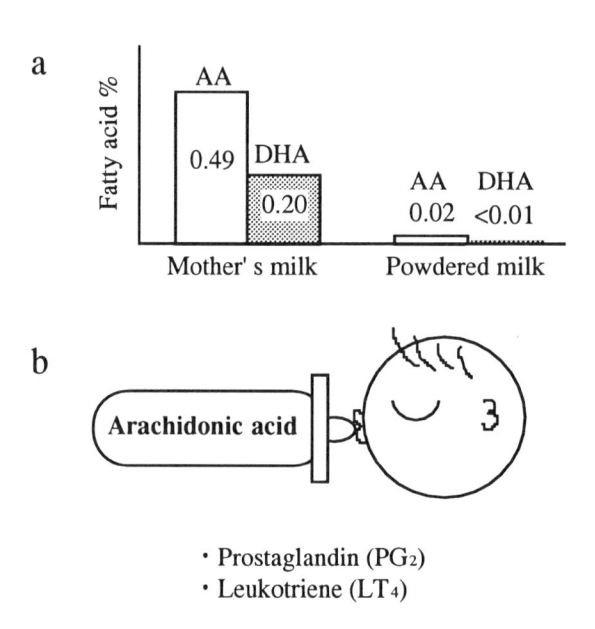

- Prostaglandin (PG$_2$)
- Leukotriene (LT$_4$)

Figure 14. a. Arachidonic acid (AA) and docosahexaenoic acid (DHA) content of mother's milk and powdered milk.
b. Function of arachidonic acid.

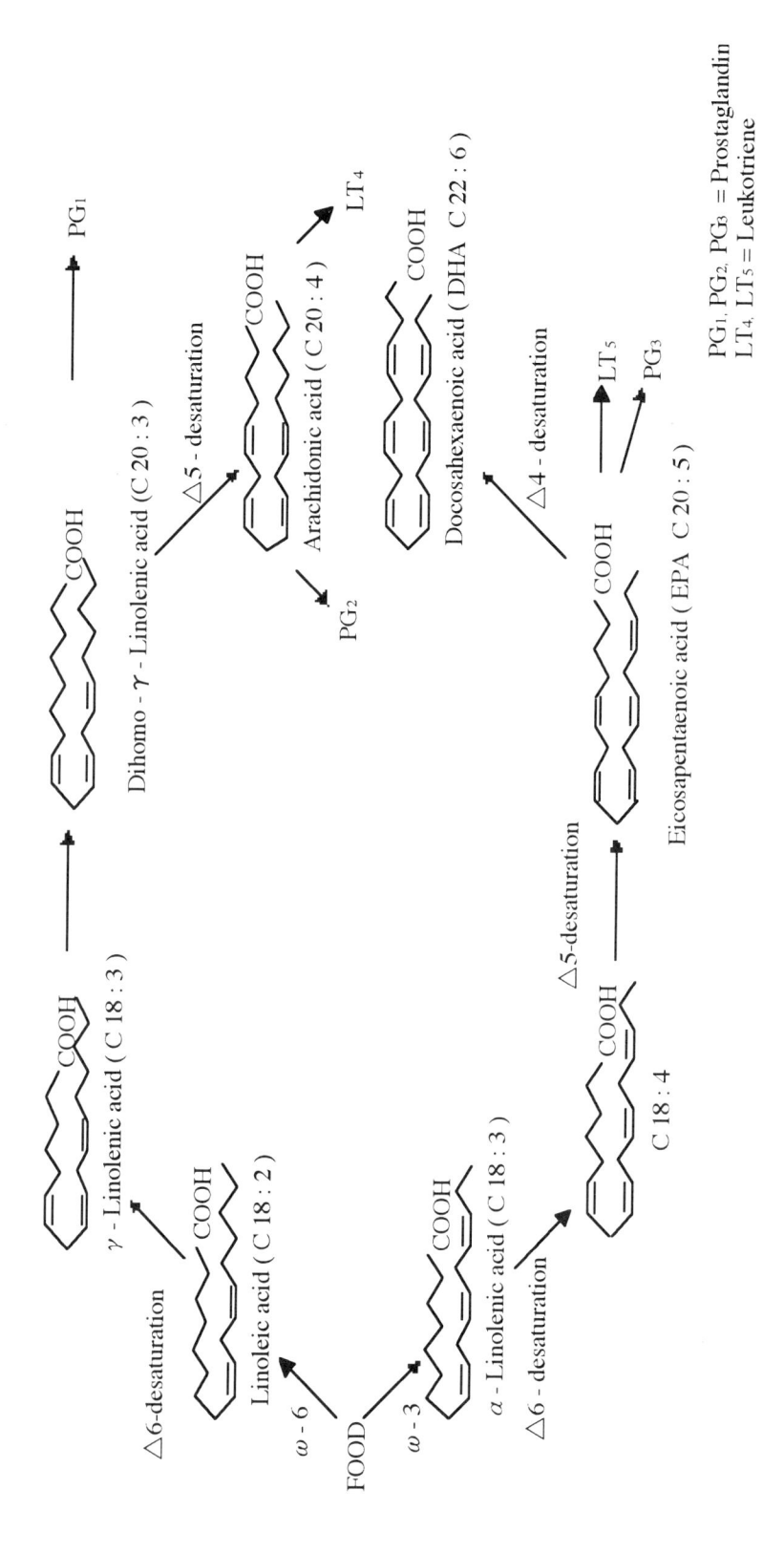

Figure 13. Metabolic pathways of ω-6 and ω-3 fatty acids.

C. EGG YOLK LIPIDS AS A SOURCE OF CHOLESTEROL

EYL also have cholesterol which is an important component in cell membranes and is needed for the rapid growth of infants. Cholesterol is a precursor of bile acids, sex hormones, and cortex hormones.

A number of commercial canned formulas based on nonfat dry milk solids have shown 5-7mg of cholesterol per 100 ml. An optimum addition of EYL having 3-5 mg/100ml cholesterol should bring commercial powdered milk close to mother's milk (Figure 15).

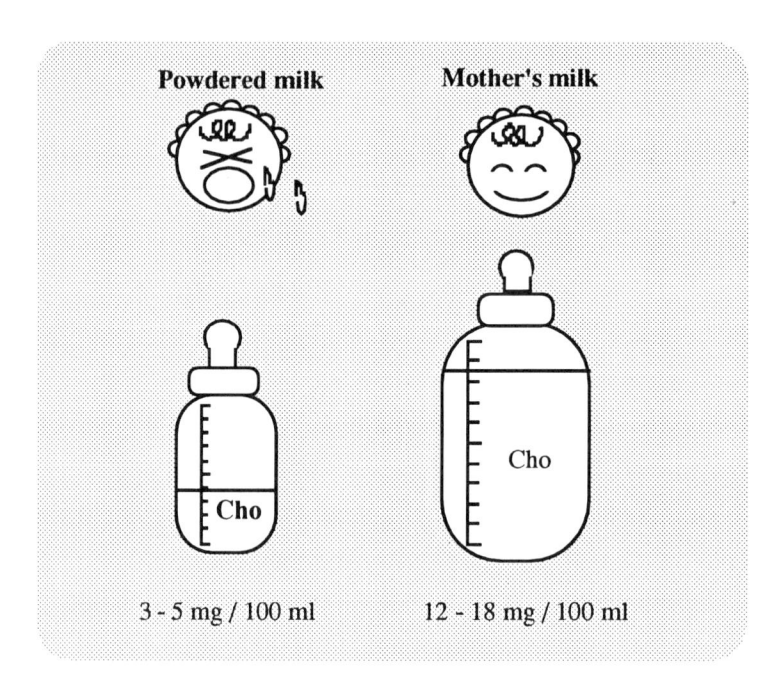

Figure 15. Cholesterol content of powdered milk and mother's milk.

D. EGG YOLK LIPIDS AS AN ANTIOXIDANT

Lecithin has usually been considered as an antioxidative synergist, improving or prolonging the action of primary antioxidants like tocopherols. The ability of PLs to enhance the antioxidative properties of primary antioxidants appear to vary with the type of oil being stabilized [83]. Dziedic has reported that PLs or their breakdown products regenerate primary antioxidants by donating hydrogen radicals or protons [84].

Antioxidant properties of PLs have been demonstrated in fish oil, vegetable oil, and animal oil.

The antioxidative properties of EYL in DHA oil were determined by the active oxygen method (AOM, Kuromachi Kagaku Co., Ltd, Tokyo, Japan). EYL (0.1%) having varying lipid content were added into 20 ml of DHA oil. Air flow in the AOM was at the rate of 1.23 ml/min, which was heated at 98°C. The oxidative stability of the oil was measured by its peroxide value [85].

The antioxidative ability of a 1.0% addition of each yolk lipid, i.e., EYL-N, EYL-30, EYL-65, and EYL-95 in DHA oil, was examined. EYL-N had no antioxidative effect, whereas EYL-95 was the most effective among the lipids tested (Figure 16). The order of antioxidative activity increases with the increase in PLs content. The antioxidative activity of yolk lipids was superior compared to soy lipids. Therefore, the antioxidative ability was dependent on the PL concentration. The antioxidative activity of PLs decreased with their respective hydrogenation.

The antioxidative activity of EYL-30 was confirmed on squalene which is susceptible to oxidation. EYL-30 showed concentration-dependent antioxidative effects (Figure 17). Therefore, EYL has an excellent antioxidative activity against fish oil and squalene which are prone to oxidation.

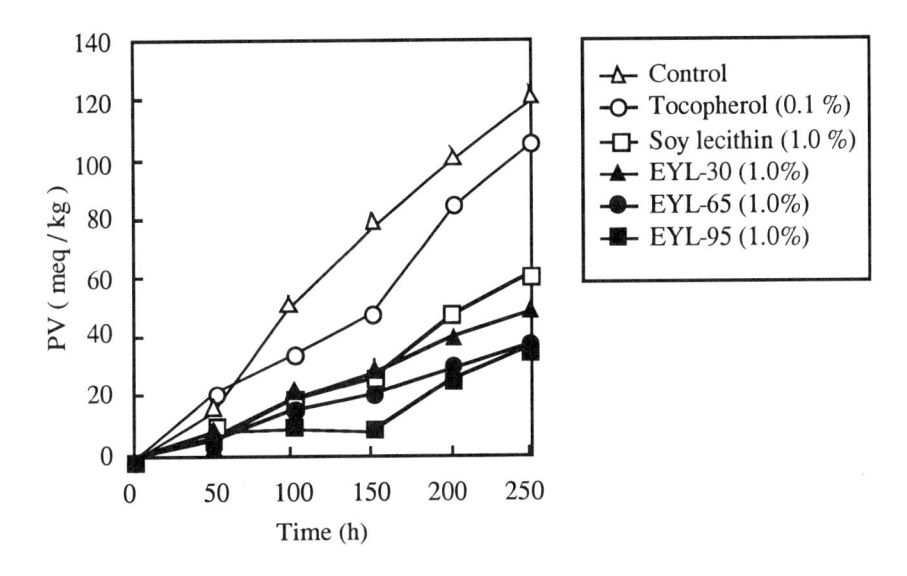

Figure 16. Antioxidative activity of egg yolk lipids, tocopherol, and soy lecithin.

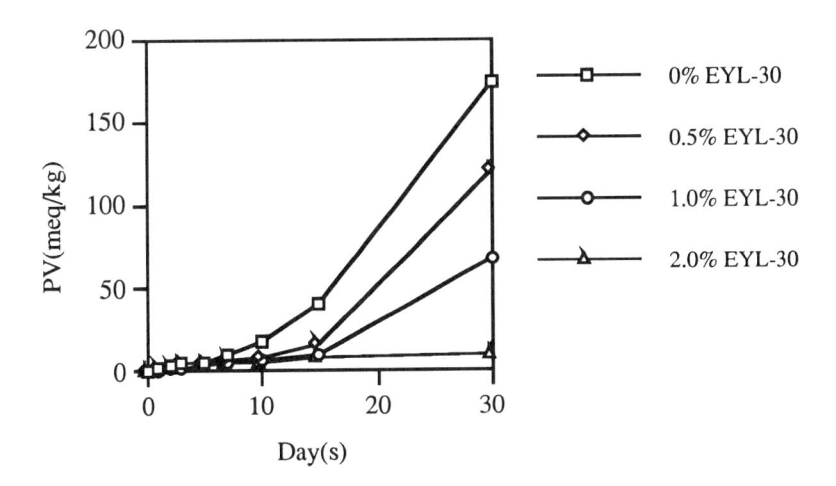

Figure 17. Antioxidative activity of various concentrations of egg yolk lipids on squalene.

E. EGG YOLK LIPIDS AS A CONSTITUENT OF LIPOSOME (PHARMACEUTICALS)

Liposome is defined as a structure consisting of one or more concentric spheres of lipid bilayers separated by water or aqueous buffer components. EYL are desirable starting materials for the manufacture of liposomes and liposomes based drug products. Partial and complete hydrogenation of yolk PLs promises to further increase their utility. These lipid spheres may offer a novel way to deliver medicines to diseased tissues. Instead of being diluted in the blood, drugs entrapped in the vesicles would reach target sites in concentrated doses [86].

EYL have been popularly used in the manufacture of pharmaceutical emulsions [87]. Research into the structure and function of biological membranes and the development and application of liposomes has led to increasing use of PLs in recent years [88].

One can prepare liposomes whose characteristics fit well for various purposes by mixing PLs in their appropriate ratios on an industrial scale. Liposomes are being studied as carriers and sustained release vehicles for vaccines, enzymes, and drugs with particular emphasis on antitumor drugs, cell modifying compounds, and hormones, etc. to be used orally or intravenously [88]. Liposomes can gather at the specific joints which cause pain in arthritic patients. A decrease in blood glucose has been found in diabetic rats fed liposomes containing insulin [89]. Liposome-encapsulated drugs have been tested for anticancer activity, for example, doxorubicin (adriamycin), and encapsulation within liposomes is very promising. Mice with tumors can tolerate about 10 times the normal lethal level of this potent but dangerous drug. A water-insoluble drug has been transported to leukemia cells using liposomes [86].

To understand the measured effects, it is of interest to take into account the dynamic shape of the molecules. The molecular shape of PLs is an important consideration in liposome modeling. Hydrated PC with its approximate cylindrical shape forms lamellar structures and LPC forms spherical micelles when dispersed in water. LPC acts as a natural modulator of biological membranes because of its inverted cone shape. The liposomes were prepared from a lipid mixture composed of PC, LPC, PA, PS, and cholesterol in various molar ratios. The liposome fractions were collected by measuring their absorbance at 660 nm. The amount of antitumor drug *cis*-diamminedichloroplatinum (CDDP) encapsulated was determined at 266 nm by atomic absorption spectrometer. A small sized unilamellar liposome (SUV) has been prepared by the combination of PC, LPC, and PA.

Usually hydrogenated PC liposomes are turbid but the liposomes containing a 30% molar ratio of LPC were transparent because of their small size (average size = 50 nm). Addition of PA or PS further improved the encapsulation efficiency, and the transparency obtained by the addition of LPC was not affected. The addition of cholesterol countered the effect of LPC, and the turbidity of liposomes was restored. The bioassay of liposomes encapsulating CDDP was carried out in human neuroblastoma cells (IMR-32). PC/ LPC/ PA liposome was found to be very effective as compared to other liposomes and free CDDP (Juneja unpublished data).

F. EGG YOLK LIPIDS IN COSMETICS

EYL are chosen for cosmetics because of their skin feel, skin absorption, emollient, spreading, dispersing, moisturizing (wetting), emulsifying and a lot of other properties [90]. EYL are used in cosmetics because of their multifunctionality. EYL are safe, nontoxic and biodegradable. EYL replenish lipid deficiency in skin and closely resemble skin lipids. Egg yolk PLs reduce the oily or greasy feeling on the skin and give an elegant feel. These PLs increase the skin absorption of the active ingredients of cosmetic formulations. The products prepared with yolk lecithin reduce transfer to clothing due to improved film adhesion [91]. EYL are potential base materials for moisturizing creams, cleansing creams, after shave lotions, hair products, bath lotions, body lotions, makeup, lipsticks, sun protective products, and gel and cream for eyes etc. EYL have been added to shampoos, rinses, skin creams, lotions, and soaps. Effects of EYL has been claimed for hair sheen and for benefits to skin, etc. [91].

Hydrogenated egg yolk PLs have been used in liposomes and other cosmetic applications. Liposomes for cosmetic usage generally are prepared by the mechanical disruption of hydrated lipids by such techniques as ultrasonication. Liposomes increase the value of cosmetic products by encapsulating and stabilizing a variety of ingredients, such as moisturizers, skin care agents, sunscreens, vitamins, and tanning agents [92].

CONCLUSION

PLs are very important compounds essential to life. They are major constituents of the cell membranes of all living organisms and are active participants in metabolic processes. As a result of their amphiphilic properties, PLs provide the necessary interface for transport systems.

Egg yolk continues to provide a food product of nearly constant composition. Controlled feeding in hens gives consistent quality of lipid products including their fatty acid spectrum compared to other lipid sources. EYL are the best quality commercial PLs. Therefore, EYL are generally used in pharmaceutical products. The temperature of isolation of EYL is below 40°C compared to soy and other lipids which are prepared at more than 80°C, leading to damage of phospholipids and their fatty acids and off-flavor of the products. The lipid researchers and manufacturers continue to strive for the production of lipid products having novel functionality. The usage of EYL will increase in the food industry because of their recent cost effective, large scale production. Some dairy industries in Europe, Japan, and other countries have added EYL as a multifunctional source of AA, DHA, and cholesterol. EYL in infant formulations should be similar to the brain growth factor found in mother's milk.

REFERENCES

1. **Juneja, L. R., Hibi, N., Yamane, T., and Shimizu, S.,** Repeated batch and continuous operations for phosphatidylglycerol synthesis from phosphatidylcholine with immobilized phospholipase D, *Appl. Microbiol. Biotechnol.*, **27**, 146, 1987.
2. **Juneja, L. R., Hibi, N., Inagaki, N., Yamane, T., and Shimizu, S.,** Comparative study on conversion of phosphatidylcholine to phosphatidylglycerol by cabbage phospholipase D in micelle and emulsion systems, *Enzyme Microb. Technol.*, **350**, 350, 1987.
3. **Juneja, L. R., Kazuoka, T., Yamane, T., and Shimizu, S.,** Kinetic evaluation of conversion of phosphatidylcholine to phosphatidylethanolamine by phospholipases D from different sources, *Biochim. Biophys. Acta*, **960**, 334, 1988.
4. **Juneja, L. R., Kazuoka, T., Goto, N., Yamane, T., and Shimizu, S.,** Conversion of phosphatidylcholine to phosphatidylserine by various phospholipases D in the presence of L- or D-serine, *Biochim. Biophys. Acta*, **1003**, 277, 1989.
5. **Juneja, L. R., Yamane, T., and Shimizu, S.,** Enzymatic method of increasing phosphatidylcholine, *J. Am. Oil Chem. Soc.*, **66**, 714, 1989.
6. **Dearden, S. J., Hunter, T. F., and Philip, J.,** A rapid method for the preparation of microvesicles of egg yolk lecithin, *Biochim. Biophys. Acta*, **689**, 415, 1982.
7. **Privette, O. S., Blank, M. L., and Schmit, J. A.,** Studies on the composition of egg lipids, *J. Food Sci.*, **27**, 463, 1962.

8. Fujino, Y., Chemistry of lipids in egg yolk, *J. Jpn. Soc. Food and Nutr.*, **24**, 317, 1971.
9. Baer, E. and Kates, M., Synthesis of enantiomeric α-lecithins, *J. Amer. Chem. Soc.*, **72**, 942, 1950.
10. Tattrie, N. H., Positional distribution of saturated and unsaturated fatty acids on egg lecithin, *J. Lipid Res.*, **1**, 60, 1959.
11. Renkonen, O., Individual molecular species of different phospholipid classes. Part II. A method of analysis, *J. Am. Oil Chem. Soc.*, **42**, 298, 1965.
12. Christie, W. W. and Moore, J. H., The structure of egg yolk triglycerides, *Biochim. Biophys. Acta*, **218**, 83, 1970.
13. Suyama, K., Adachi, S., Sugawara, H., and Honjoh, H., Stereospecific analysis of glycerolipids of egg yolk of Japanese quail (*Coturnix coturnix japonica*), *Lipids*, **14**, 707, 1979.
14. Weiner, N., Martin, F., and Riaz, M., Liposomes as a drug delivery system, *Drug dev. Ind. Pharm.*, **15**, 1523, 1989.
15. Thudichum, J. L. W., *A Treatise on the Chemical Constitution of the Brain, Balliere, Tindall and Cox*, London, 1884.
16. Noble, R. C. and Moore, J. H., Metabolism of the yolk phospholipids by the developing chick embryo, *Can. J. Biochem.*, **43**, 1677, 1965.
17. Fujino, Y., Negishi, T., Momma, H., and Yamabuki, S., Studies on the lipids of egg yolk. Part II. Nature of sphingolipids, *Agric. Bio. Chem.*, **35**, 134, 1971.
18. Do, U. H., Pei, P. T., and Minard, R. D., Separation of molecular species of ceramides as benzoyl and p-nitrobenzoyl derivatives by high performance liquid chromatography, *Lipids*, **16**, 855, 1981.
19. Connelly, P. W. and Kuksis, A., Influence of divalent cations on rat apolipoprotein transfer to synthetic lipoprotein like lipid emulsions *in vitro*, *Can. J. Biochem. Cell. Biol.*, **61**, 63, 1983.
20. Vargas, R. E., Allred, J. B., Biggert, M. D., and Naber, E. C., Effect of dietary 7-ketocholesterol, pure, or oxidized cholesterol on hepatic 3-hydroxy-3-methylglutaryl-coenzyme A reductase activity, energy balance, egg cholesterol concentration, and ^{14}C-acetate incorporation into yolk lipids of laying hens, *Poult. Sci.*, **65**, 1333, 1986.
21. Kodchodkar, B. J., Horlick, L., and O'Neil, J. B., Absorption of dietary β-sitosterol in laying hens and its incorporation into the egg, *J. Nutr.*, **106**, 1629, 1976.
22. Fenton, M. and Sim, J. S., Determination of egg yolk cholesterol content by on-column capillary gas chromatography, *J. Chromatogr.*, **540**, 323, 1991.
23. Li, S.-C., Chien, J.-L., Wan, C.-C., and Li, Y.-T., Occurrence of glycosphingolipids in chicken egg yolk, *Biochem. J.*, **173**, 697, 1978.
24. Scholfield, C. R., Occurrence, structure, composition and nomenclature, in *Lecithins*, Szuhaj, B. F. and List, G.R., Eds., American Oil Chemists' Society, Champaign, Illinois, 1985, p 1.
25. Van Nieuwenhuyzen, W., Lecithin production and properties, *J. Am. Oil Chem. Soc.*, **53**, 425, 1976.
26. Ansell, G. B. and Spanner, S., Phosphatidylserine, phosphatidylethanolamine and phosphatidylcholine, in *Phospholipids (New Comprehensive Biochemistry Vol. 4)*, Hawthorn, J. N., Ansell, G.B., Eds., Elsevier, Amsterdam, 1982, p 1.
27. Pangborn, M. C., A simplified purification of lecithin, *J. Biol. Chem.*, **188**, 471, 1950.
28. Lundberg, B., Isolation and characterization of egg lecithin, *Acta Chem. Scand*, **27**, 2515, 1973.
29. Yano, N., Fukinbara, I., and Takano, M., A process for obtaining yolk lecithin from, raw egg yolk, *US patent* 4157404, 1979.

30. Aneja, R., Chadha, J. S., and Yoell, R. W., A process for the separation of phosphatide mixtures: The preparation of phosphatidylethanolamine-free phosphatides from soy lecithin, *Fette-seifen-Anstrichmittel*, **73**, 643, 1971.
31. Glass, R. L., Separation of phospholipid and its molecular species by high performance liquid chromatography, *J. Agric. Food Chem*, **38**, 1684, 1990.
32. Kiyashchitsky, B. A., Mezhova, I. V., Krasnopoisky, Y.-M., and Shvets, V. I., Preparative isolation of polyphosphoinositides and other anionic phospholipids from natural sources using chromatography on adsorbents containing primary amino groups, *Biotechnol. Appl. Biochem.*, **14**, 284, 1991.
33. Hanson, V. L., Park, J. Y., Osborn, T. W., and Kiral, R. M., High performance liquid chromatographic analysis of egg yolk phospholipids, *J. Chromatogr.*, **205**, 393, 1981.
34. Singleton, W. S., Gray, M. S., Brown, M. L., and White, J. L., Chromatographically homogeneous lecithin from egg phospholipids, *J. Am. Oil Chem. Soc.*, **42**, 53, 1965.
35. Hanras, C. and Perrin, J. L., Gram-scale preparative HPLC of phospholipids from soybean lecithins, *J. Am. Oil Chem. Soc.*, **68**, 804, 1991.
36. Chen, S. S. and Kou, A. Y., Improved procedure for the separation of phospholipids by high performance chromatography, *J. Chromatogr.*, **227**, 25, 1982.
37. Geurts van Kessel, W. S. M., Tieman, M., and Demel, R. A., Purification of phospholipids by preparative high pressure liquid chromatography, *Lipids*, **16**, 58, 1981.
38. Fager, R. S., Shapiro, S., and Litman, B. J., A large scale preparation of phosphatidylethanolamine, lysophosphatidylethanolamine and phosphatidylcholine high performance liquid chromatography: a partial purification of molecular species, *J. Lipid Res*, **18**, 704, 1977.
39. Gunther, B. R., Method for the preparation of phosphatidylcholine of low oil content, *US patent* 4496486, 1985.
40. Porter, N. A., Wolf, R. A., and Nixon, J. R., Separation and purification of lecithin by high performance liquid chromatography, *Lipids*, **14**, 20, 1978.
41. Do, U. H. and Ramachandran, S., Mild alkali-stable phospholipids in chicken egg yolk: Characterization of 1-alkenyl and 1-alkyl-sn-glycero-3-phosphoethanolamine, sphingomyelin, and 1-alkyl-sn-glycero-3-phosphocholine., *J. Lipid Res.*, **21**, 888, 1980.
42. Ramesh, B., Prabhudesai, A. U., and Vishwanathan, C. V., Simultaneous extraction and preparative fractionation of egg yolk lipids using the principle of adsorption, *J. Am. Oil Chem. Soc.*, **55**, 501, 1978.
43. Szuhaj, B. F., Lecithin production and utilization, *J. Am. Oil Chem. Soc.*, **60**, 306, 1983.
44. Baharami, S., Gasser, H., and Redl, H., A preparative high performance liquid chromatography method for the separation of lecithin: Comparison to thin layer chromatography, *J. Lipid Res.*, **28**, 596, 1987.
45. Kolarovic, L. and Fournier, N. C., A comparison of extraction methods for the isolation of phospholipids from biological sources, *Anal. Biochem.*, **156**, 244, 1986.
46. Nielsein, J. R., A simple chromatographic method for purification of egg lecithin, *Lipids*, **15**, 481, 1980.
47. Primes, K. J., Sanchez, R. A., Metzner, E. K., and Patel, K. M., Large scale purification of Phosphatidylcholine from egg yolk phospholipids by column chromatography, *J. Chromatogr.*, **236**, 519, 1982.
48. Amari, J. V., Brown, P. R., Grill, C. M., and Turootte, J. G., Isolation and purification of lecithin by preparative high-performance liquid chromatography, *J. Chromatogr.*, **517**, 219, 1990.
49. Farag, R. S., El-Sharabassy, A. A. M., Abdel Rahim, G. A., Hewedy, E. M., and Ragab, A. A., Biochemical studies on phospholipids of hen's egg during incubation, *Steifen-*

ote-Fette-wachse, **110**, 122, 1984.
50. **Takamura, H. and Kito, M.**, A highly sensitive method for quantitative analysis of phospholipid molecular species by high performance liquid chromatography, *J. Biochem*, **109**, 436, 1991.
51. **Fronning, G. W., Wehling, R. L., Cuppett, S. L., Pierce, M. M., Niemann, L., and Siekman, D. K.**, Extraction of cholesterol and other lipids from dried egg yolk using supercritical carbon dioxide, *J. Food Sci.*, **55**, 95, 1990.
52. **Juneja, L. R., Sugino, H., Fujiki, M., Kim, M., and Yamamoto, T.**, Preparation of pure phospholipids from egg yolk, in *Egg Uses and Processing Technologies - New Developments*, Sim, J. S. and Nakai, S., Eds., CAB International, Oxon, 1994, p 139.
53. **Juneja, L. R.**, *"Enzymatic conversion of polar groups of phospholipids."* Ph. D. Thesis, Nagoya University, 1989.
54. **Owens, K.**, A two-dimensional thin-layer chromatographic procedure for the estimation of plasmalogens, *Biochem. J.*, **100**, 354, 1966.
55. **Patton, G. M., Fasulo, J. M., and Robins, S. J.**, Separation of phospholipids and individual molecular species of phospholipids by high-performance liquid chromatography, *J. Lipid Res.*, **23**, 190, 1982.
56. **Dugan, L. L., Demediuk, P., Pendley, C. E. I., and Horrocks, L. A.**, Separation of phospholipids by high-performance liquid chromatography: all major classes, including ethanolamine and choline plasmalogens, and most minor classes, including lysophosphatidylethanolamine, *J. Chromatogr.*, **378**, 317, 1986.
57. **Schlenk, H. and Gellerman, J. L.**, Esterfication of fatty acids with diazomethane on a small scale, *Anal. Chem.*, **32**, 1412, 1960.
58. **Schmid, P. C., Natrajan, V., Weis, B. K., and Schmid, H. H.**, Hydrolysis of N-acylated glycerophospholipids by phospholipases A2 and D: A method of identification and analysis, *Chem. Phys. Lipids*, **41**, 195, 1986.
59. **Yoo, I.-J.**, Innovative egg products and future trends in Korea, in *Egg Uses and Processing Technologies, New Developments*, Sim, J. S. and Nakai, S., Eds., CAB International, Oxon, 1994, p 63.
60. **Jiang, Z. and Sim, J. S.**, Fatty acid modification of yolk lipids and cholesterol lowering eggs, in *Egg Uses Processing Technologies, New Developments*, Sim, J. S. and Nakai, S., Eds., CAB International, Oxon, 1994, p 349.
61. **Gilbert, A. B.**, *Physiology and Biochemistry of the Domestic Fowl*, Bell, D. J. and Freeman, B. M., Eds., Academic Press, New York, 1971.
62. **Naber, E. C.**, The cholesterol problem, the egg and lipid metabolism in the laying hen, *Poult. Sci.*, **55**, 14, 1976.
63. **Prosise, W. E.**, Commercial lecithin products: Food use of soybean lecithin, in *Lecithins*, Szuhaj, B. F. and List, G.R., Eds., American Oil Chemist's Society, Champaign, Illinois, 1985, p 163.
64. **Hirsch, M. J. and Wurtman, R. J.**, Lecithin consumption increases acetylcholine concentrations in rat brain and adrenal gland, *Science*, **202**, 223, 1978.
65. **Zeisel, S. H.**, Dietary choline: Biochemistry, physiology and pharmacology, *Ann. Rev. Nutr.*, **1**, 95, 1981.
66. **Zeisel, S. H. and Canty D. J.**, Choline phospholipids: Molecular mechanisms for human disease: A meeting report, *J. Nutr. Biochem.*, **4**, 258, 1993.
67. **Zeisel, S. H., Da Costa, K.-A., Franklin, P. D., Alexander, E. A., Lamont, J. T., Sheard, N. F., and Beiser, A.**, Choline, an essential nutrient for humans, *FASEB J.*, **5**, 2093, 1991.
68. **Lieber, C., Robins, S. J., Li, J., Decarli, L. M., Mak, K. M., Fasulo, J.-M., and Leo, M. A.**, Phosphatidylcholine protects against fibrosis and cirrhosis in the baboon,

Gastroenterology, **106**, 152, 1994.
69. **Best, C. H. and Huntsman, M. E.**, The effects of the components of lecithine upon deposition of fat in the liver, *J. Physiol.*, **75**, 405, 1932.
70. **Michael, U. F., Cookson, S. L., Chavez, R., and Pardo, V.**, Renal function in the choline deficient rat, *Proc. Soc. Exp. Biol. Med.*, **150**, 672, 1975.
71. **Griffith, W. H. and Wade, N. J.**, Choline metabolism I. The occurrence and prevention of hemorrhagic degeneration in young rats on a low choline diet, *J. Biol. Chem.*, **131**, 567, 1939.
72. **Patterson, J. M. and McHenry, E. W.**, Choline and the prevention of hemorrhagic kidneys in the rat, *J. Biol. Chem.*, **156**, 265, 1944.
73. **Chang, C. H. and Jensen, L. S.**, Inefficacy of carnitine as a substitute for choline for normal reproduction in Japanese quail, *Poult. Sci.*, **54**, 1718, 1975.
74. **Jukes, T. H.**, Prevention of perosis by choline, *J. Biol. Chem.*, **134**, 789, 1940.
75. **Pawelczyk, T. and Lowenstein, J. M.**, Inhibition of phospholipase delta by hexadecylphosphoryl choline and lysophospholipids with antitumor activity, *Biochem. Pharmacol.*, **45**, 493, 1993.
76. **Jenike, M. A., Albert, M. S., Heller, H., LoCastro, S., and Gunther, J.**, Combination therapy with lecithin and ergoloid mesylates for Alzheimer's disease, *J. Clin. Psychiatry*, **47**, 249, 1986.
77. **Davies, P. and Maloney, A. J.**, Selective loss of central cholinergic neurons in Alzheimer's disease [letter], *Lancet*, **2**, 1403, 1976.
78. **Neuringer, M., Connor, W. E., Van Petten, C., and Barstad, L.**, Dietary ω-3 fatty acid deficiency and visual loss in infant rhesus monkeys, *J. Clin. Invest.*, **73**, 273, 1984.
79. **Innis, S. M.**, Sources of ω-3 fatty acids in Arctic diets and their effect on red cell and breast milk fatty acids in Canadian Inuit, in *Dietary ω-3 and ω-6 Fatty Acids-Biological Effects and Nutritional Essentiality*, Galli, C. and Simopoulos, A. P., Eds., Plenum Press, New York, 1989, p 145.
80. **Carlson, S. E.**, Dietary fatty acid in relation to neural developments in humans, in *Dietary ω-3 and ω-6 Fatty Acids-Biological Effects and Nutritional Essentiality*, Galli, S. and Simopoulos, A. P., Eds., Plenum Press, New York, 1989, p 135.
81. **Carlson, S. E., Rhodes, P. G., and Ferguson, M. G.**, Docosahexaenoic acid status of preterm infants at birth and following feeding with human milk or formula, *Am. J. Clin. Nutr.*, **44**, 798, 1986.
82. **Chambaz, J., Ravel, D., Manier, M. C., Pepin, D., Mulliez, N., and Bereziat, G.**, Essential fatty acids interconversion in the human fetal liver, *Biol. Neonate*, **47**, 136, 1985.
83. **Dziedzic, S. Z. and Hudson, B. J. F.**, Phosphatidylethanolamine as a synergist for primary antioxidants in edible oils, *J. Am. Oil Chem. Soc.*, **61**, 1042, 1984.
84. **Dziedzic, S. Z.**, Fate of propyl gallate and diphosphatidylethanolamine in lard during autoxidation at 120°C, *J. Agric. Food Chem.*, **34**, 1027, 1986.
85. **Yamamoto, Y. and Omori, M.**, Antioxidative activity of egg yolk lipoproteins, *Biosci. Biotech. Biochem.*, **58**, 1711, 1994.
86. **Ostro, M. J.**, Liposomes, *Sci. Am.*, **256**, 103, 1987.
87. **Baker, M. E.**, Invertebrate vitellogenin is homologous to human von Willebrand factor [letter], *Biochem. J.*, **256**, 1059, 1988.
88. **Tyrrell, D. A., Heath, T. D., Colley, C. M., and Ryman, B. E.**, New aspects of liposomes, *Biochim. Biophys. Acta*, **457**, 259, 1976.
89. **Patel, H. M. and Ryman, B. E.**, Oral administration of insulin by encapsulation within liposomes, *FEBS Letters*, **62**, 60, 1976.
90. **Baker, C.**, Lecithin in cosmetics, in *Lecithins*, Szuhaj, B. F., Ed., Am. Oil Chem. Soc.,

Champaign, Illinois, USA, 1989, p 253.
91. **Sagarin, E.,** *Cosmecis Science and Technology*, Interscience Publishers, Inc., New York, 1957.
92. **Van Biervliet, J. P., Labeur, C., and Rosseneu, M.,** Lipoprotein(α) in newborns, *Atherosclerosis*, **86**, 173, 1991.

Chapter 7

GLYCOCHEMISTRY OF HEN EGGS

M. Koketsu

TABLE OF CONTENTS

I. Introduction
II. Free Sugars of Hen Eggs
III. Oligosaccharides Conjugated with Egg Proteins
    A. Ovomucoid
    B. Ovalbumin
    C. Ovotransferrin
    D. Phosvitin
    E. Ovomucin
    F. Yolk Riboflavin-binding Protein
    G. Egg-yolk Immunoglobulin
IV. Glycolipids of Hen Eggs
V. Sialic Acid in Hen Eggs
    A. Distribution of Sialic Acid in Various Fractions of Hen Eggs
    B. Isolation of Sialic Acid from Egg Yolk
    C. Sialyloligosaccharides from Egg Yolk
V. Biological Activities of Egg Yolk Sialyloligosaccharides
    A. Rotaviral inhibitory Effect of Egg Yolk Sialyloligosaccharides
    B. Learning Performance of Sialyloligosaccharides

References

I. INTRODUCTION

Avian egg contains several kinds of free sugars and glycoconjugates. Glucose is the most common of the free sugars. The glycoconjugates are glycoproteins and glycolipids, and the structures of glycoproteins found in both egg yolk and egg white have been extensively studied. Most of the glyco moieties of glycoproteins are *N*-linked and/or *O*-linked sugar chains. The sugar chains in the glycoproteins play several important physiological and physicochemical parts in cells of newborn chicken and in eggs [1]. The structures of glycolipids in eggs have also been reported [2].

Sialic acid is a functional carbohydrate molecule [3]. It is known to exist in all kinds of animal cells and certain microbes (*Escherichia coli, Neisseria meningitidis* and *Salmonella* strains) [4-6]. Hen eggs contain sialic acid in such amounts that they are thought to be a potential source for the isolation of sialic acid on an industrial scale [7, 8].

In this chapter, carbohydrates in hen eggs and the structures of the oligosaccharides of the major egg glycoproteins, together with the functions of egg yolk sialyloligosaccharides, are described.

II. FREE SUGARS OF HEN EGGS

Glucose is a major component of free sugars found in hen eggs, and its concentrations in whole egg, albumen and yolk are 0.7, 0.8, and 0.7 %, respectively [9]. Free glucose causes the maillard reaction changing the color of egg white to brown on heating. Therefore, before drying egg white in hot air, the glucose is usually removed by fermentation with yeast [10] or by enzymatic oxidation.

III. OLIGOSACCHARIDES CONJUGATED WITH EGG PROTEINS

Most of the proteins in hen eggs are glycoderivatives. Their sugar chains are classified into several types in the structure but with some microheterogeneity. As is known with various animal proteins, the glycoproteins in eggs involve two types of sugar chain-protein link: the N-linked type formed through the β-amino group of the asparagine residue and the O-linked type formed through the hydroxyl group of serine or the threonine residue. The structures of sugar chains of the major glycoproteins in hen eggs are described in this cahapter.

A. OVOMUCOID

Ovomucoid is a typical glycoprotein with a molecular weight of 2.8×10^4 and it is about 11.0 % of total egg white protein [11]. The saccharides of ovomucoid link through asparagine unlike mucin-type sugar chains, and the saccharide moiety occupies 25-30 % of the whole molecule. Ovomucoid has four major binding sites with sugar chains. The glycosylated Asn residues have been reported to locate at positions 10, 53, 69, and 75 [12, 13]. The oligosaccharides of ovomucoid are neutral- [14-20], sialyl- and sulfated oligosaccharides [16, 19] with microheterogeneity in their structures. The structures of neutral oligosaccharides in ovomucoid are shown in Figure 1.

B. OVALBUMIN

Ovalbumin makes up about 54.0% of total egg white protein. It is also a major phosphoglycoprotein of egg white proteins. Its molecular weight is approximately 4.5×10^4 [21] containing 3.2 % sugars. The major site of ovalbumin binding of carbohydrates is at position 292 [22]. The sugar chain structures are shown in Figure 2 [19, 23-28]. The sugar chains of ovalbumin are known to be N-linked complex and high mannose types.

C. OVOTRANSFERRIN

Ovotransferrin is a glycoprotein with a molecular weight of about 7.7×10^4 [29] and makes up about 12.0 % of total egg white protein. Ovotransferrin has only one oligosaccharide moiety binding the Asn located at position 473 [29]. The structure of the oligosaccharide is shown in Figure 3 [30, 31]. It is triantennary with bisecting GlcNAc and is devoid of galactose and sialic acid as constituents.

D. PHOSVITIN

Phosvitin is a phosphoglycoprotein with an approximate molecular weight of 2.8-3.4×10^4 [32, 33]. It contains 6.5% of carbohydrate and makes up approximately 11 % of total egg yolk protein. The major site of phosvitin binding to carbohydrate is the Asn residue located at position 169 [34]. The structures are shown in Figure 4 [34-36]. The major structure of the oligosaccharide of phosvitin is triantennary, and two of its sugar chains are terminated with sialic acid residues [34].

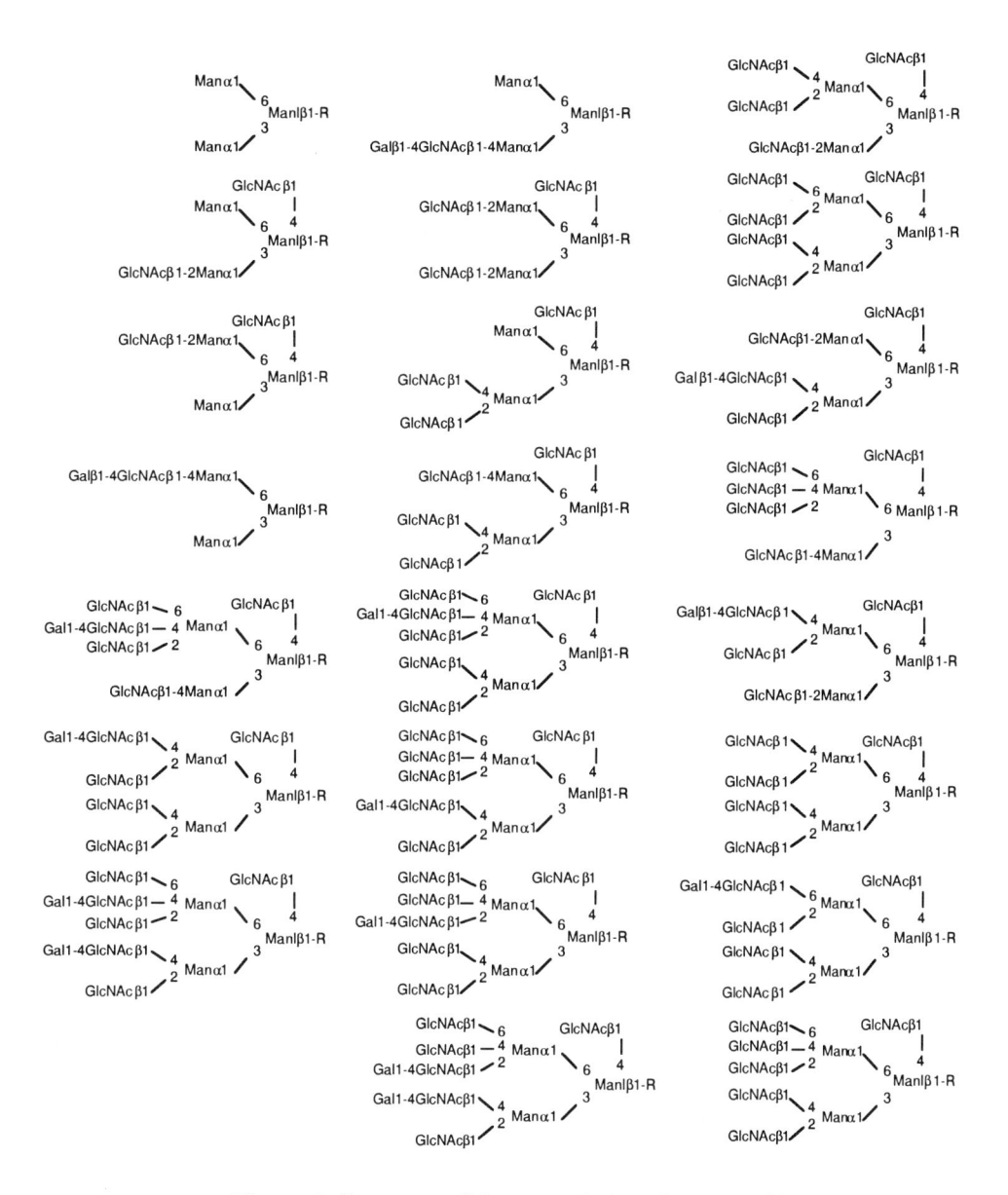

Figure 1. Structures of the sugar chains of ovomucoid.
R: 4GlcNAcβ1-4GlcNAc.

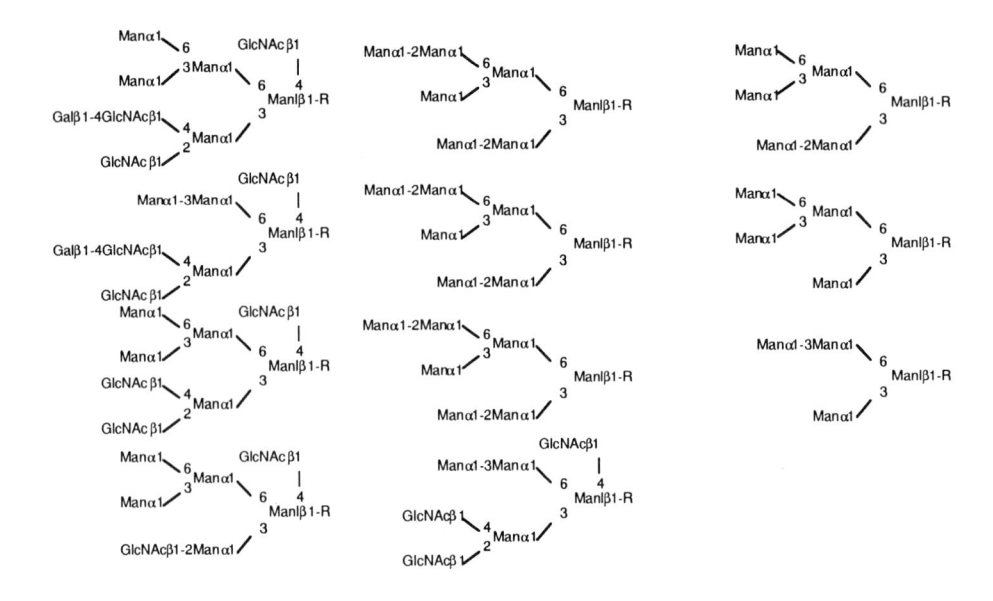

Figure 2. Structures of the sugar chains of ovalbumin.
R: 4GlcNAcβ1-4GlcNAc.

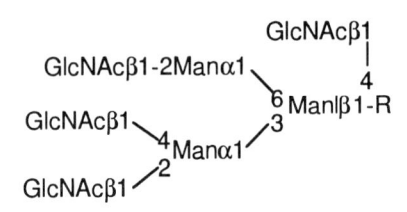

Figure 3. Structure of the sugar chain of ovotransferrin.
R: 4GlcNAcβ1-4GlcNAc.

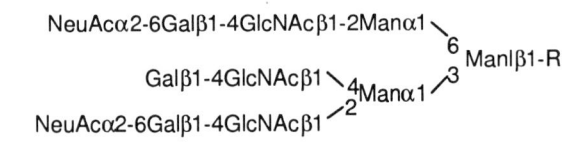

Figure 4. Structure of the sugar chain of phosvitin.
R: 4GlcNAcβ1-4GlcNAc.

E. OVOMUCIN

Ovomucin is a glycoprotein with a network structure which makes egg white a thick fluid. The molecular weight is estimated to be 1.1×10^5 [37, 38] and it makes up about 3.5% of total egg white protein. The oligosaccharides of ovomucin are linked to serine and threonine through their hydroxy groups. The oligosaccharides contain sulfated and/or sialylated residues, such as *N*-acetylneuraminyl-(2–3)-galactosyl-(1–3)-*N*-acetylgalactosaminitol-6-sulfate [39-43].

F. YOLK RIBOFLAVIN-BINDING PROTEIN

The yolk riboflavin-binding protein, a yellow phosphoglycoprotein, has an approximate molecular weight of 3.6×10^4 [44-46]. Yolk riboflavin-binding protein has two sites bound with sugar chains. The glycosylated Asn residues are located at positions 36 and 147 [47, 48]. As shown in Figure 5 [49], all the oligosaccharides have sialic acid at their non reducing ends and their structures are a complex type.

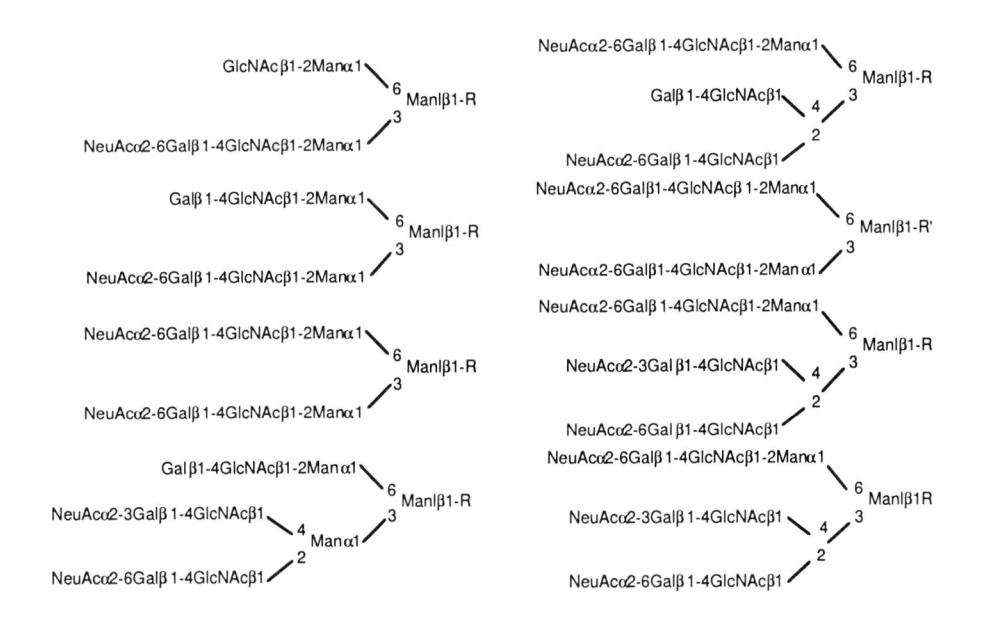

Figure 5. Structures of the sugar chains of yolk rivoflavin-binding protein. R: 4GlcNAcβ1-4GlcNAc, R': 4GlcNAcβ1-4(Fucα1-6)GlcNAc.

G. EGG-YOLK IMMUNOGLOBULIN

Egg-yolk immunoglobulin is a glycoprotein with an approximate molecular weight of 1.8×10^5 and makes up about 5 % of total egg yolk protein. Egg-yolk immunoglobulin (IgY), formerly classified as γ–livetin, has four major sites to bind oligosaccharides. The glycosylated Asn residues are at position 10, and a certain residue is between 49 and 60, and at 69 and 75. The structure of the various sugar chains of IgY indicated that 27.1% of asparagine-linked carbohydrate chains of IgY have glucose as the nonreducing end residue of the glycolic chains. The sugar chains of IgY are very unique because such glycolic chains have not been found for IgG. The sugar chain structures are shown in Figure 6 [50].

Figure 6. Structures of the sugar chains of yolk immunogloblin.
R: 4GlcNAcβ1-4GlcNAc, R': 4GlcNAcβ1-4(Fucα1-6)GlcNAc.

IV. GLYCOLIPIDS OF HEN EGGS

Li and his colleagues studied the structure of glycolipids in egg yolk [2]. They isolated galactosyl ceramide, N-acetylneuraminosylgalactosylceramide (GM4), N-acetylneuraminosyl-lactosylceramide (GM3), and di-4N-acetylneuraminosyl-lactosylceramide (GD3) with yields of 7.5 mg, 2.5 mg, 8.5 mg, and 1.5 mg, respectively, from the yolks of twelve eggs.

V. SIALIC ACID IN HEN EGGS

Sialic acid as a component of glycoconjugates in glycoproteins and glycolipids is also found in several cytological fractions. Recently, some of the roles played by sialic acid in the biological events in cells have been elucidated. More than 30 kinds of sialic acid derivatives have so far been isolated from various sources [3] and fourteen kinds of sialyl derivatives were found in the bovine submandibular gland. On the other hand, the sialic acid in hen eggs is only N-acetylneuraminic acid [2, 8]. This fact suggests that hen eggs are an excellent source of N-acetylneuraminic acid or its derivatives. The isolation of N-acetylneuraminic acid or its oligosaccharide derivatives was, thus, undertaken by our research group [7, 8].

A. DISTRIBUTION OF SIALIC ACID IN VARIOUS FRACTIONS OF HEN EGGS

The quantitative analysis of sialic acid in various fractions of hen eggs (fresh eggs) revealed that sialic acid is distributed in almost all the structural fractions of eggs. The study also made it clear that egg yolk has the highest amount of sialic acid among the various structural fractions, though the concentration of sialic acid in each fraction was highest in egg yolk membrane and chalaza as shown in Table 1 [7]. The content of sialic acid was estimated to be 0.95 g per kg of fresh egg yolk.

Table 1

Sialic Acid Content in Several Fractions of Hen Egg

Fractions	Qty. (kg/ton egg wet wt. basis)	Sialic acid yield (g)	yield (% as dry matter)
Eggshell	104.8	2.98	0.004
Shell membrane	6.2	1.22	0.07
Egg white	603.5	60.35	0.10
Egg yolk	281.0	267.0	0.19
Egg yolk membrane	2.3	3.52	1.80
Chalaza	2.2	3.96	2.40

Reproduced from Juneja et al. (1991) [7] by permission of Elsevier Science Ltd., Oxford.

B. ISOLATION OF SIALIC ACID FROM EGG YOLK

Egg yolk powder weighing 4.0 kg, obtained by a spraying method, was homogenized with three weights of ethanol. After filtration of the mixture, the residue (2.0 kg delipidated egg yolk; DEY) was homogenized with three weights of water, and the slurry was acidified to pH 1.4 with 3 M H_2SO_4 and heated at 80°C for 1 h. After cooling, the slurry was neutralized with saturated $Ba(OH)_2$ to attain a pH of 6.0, and the filtrate of the mixture was subjected to electrodialysis. The desalinate, thus, obtained was then chromatographed on a column of Dowex HCR-W2, H form (1.5 × 30 cm, 20-50 mesh), followed by a column of Dowex 1×8, formate form (1.5× 30 cm, 200-400 mesh). The sialic acid adsorbed on the latter column was eluted with a linear gradient of formic acid from 0 to 2 N at a flow rate of 2 ml/min. The sialic acid fractions were collected, evaporated at 45 °C under reduced pressure, the residue was decolored with activated charcoal and finally lyophilized. The yield of NeuAc from 2 kg of the DEY was about 2 g and its purity was >97 % (TBA method). The results analyzed by TLC, ^{1}H-, ^{13}C-NMR (Figure 7), and IR revealed that the sialic acid obtained was only N-acetylneuraminic acid. The sialic acid isolated from several other fractions of hen eggs were also N-acetylneuraminic acid alone.

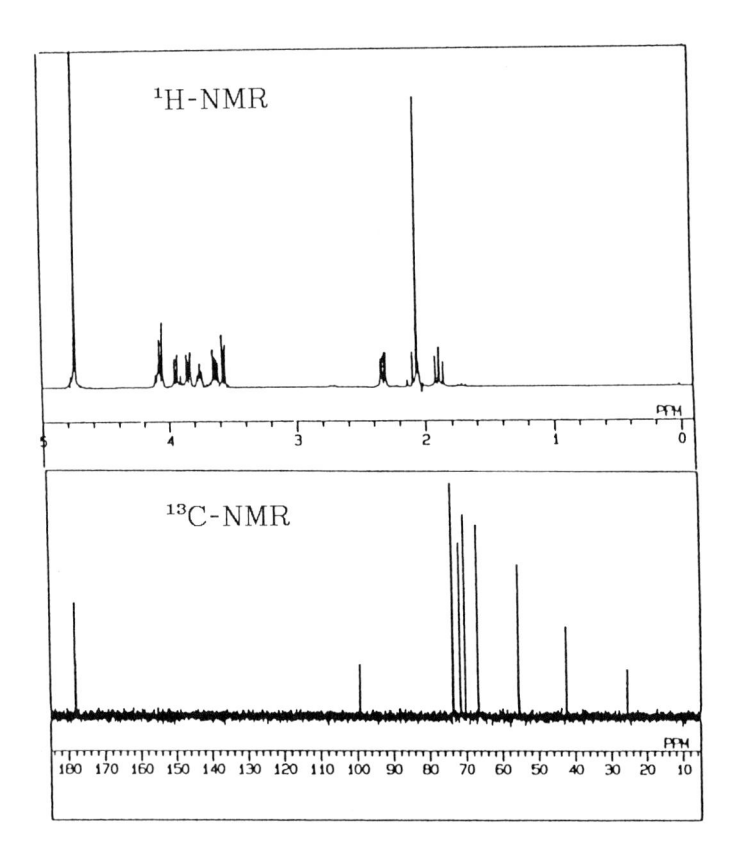

Figure 7. ^{1}H- and ^{13}C-NMR spectra of N-acetylneuraminic acid.
(Solvent, D_2O; internal standard, sodium 4,4-dimethyl-4-silapentane-1-sulfonate)
Reproduced from Koketsu et al. (1992) [8] by permission of Chapman & Hall, London.

C. SIALYLOLIGOSACCHARIDES FROM EGG YOLK

Several sialyloligosaccharides contained in egg yolk were isolated and their structures were determined [51, 52]. The supernatant of the water extract of delipidated egg yolk described above (DEY) was dialyzed using a membrane for cutting off the substances of molecular weights less than about 1,000. The asparagine-linked oligosaccharides in the concentrate were liberated by hydrazinolysis/ re N-acetylation and the reducing ends of the resulting oligosaccharides were labeled with p-aminobenzoic ethyl ester (ABEE) to be a UV-absorbing compound [53, 54]. The ABEE-derivatized oligosaccharides were fractionated by anion exchange (Figure 8) and reversed phase HPLC [55] (Figure 9). Of the total sialyloligosaccharides obtained from the water extract of DEY, monosialyl and disialyloligosaccharides were found to be 47.7 and 50.6%, respectively. The quantities of SI-1, SI-2, and SII-1 in Figure 9 were calculated to be 9.6, 6.5 and 17.5%, respectively, on the basis of UV absorbance at 304 nm. The structures of the three isolated sialyloligosaccharides were determined by analysis with HPLC and ^{1}H-NMR [56]. The major sialyloligosaccharides isolated from DEY were the N-acetyllactosamine type (Figure 10).

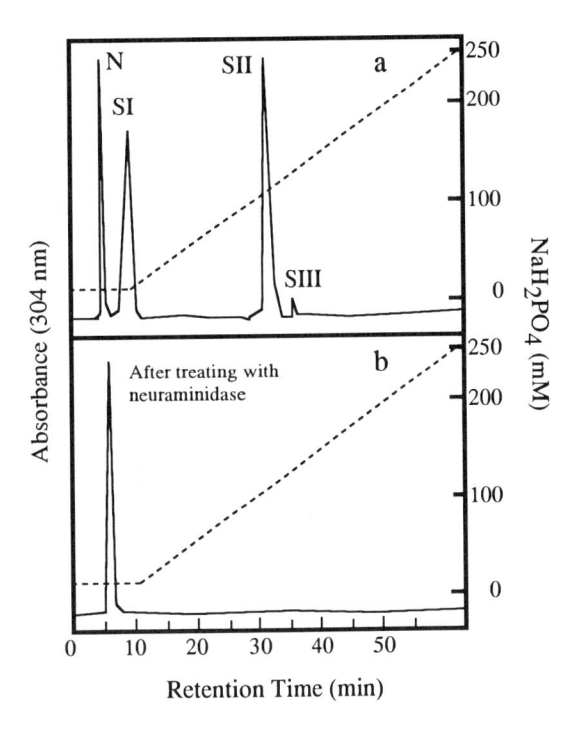

Figure 8. Separation of ABEE-sialyloligosacchrides from delipidated egg yolk on a DEAE-5PW anion exchange column (0.75 × 7.5 cm) with a linear gradient of NaH2PO4 from 10-250 mM NaH2PO4. Flow rate: 0.5 mL/min; detection: 304 nm. (a), the elution profile of acidic oligosaccharides; (b), the neuraminidase digest of (a). (N, SI, SII, and SIII indicate the elution position of standard neutral oligosaccharide, mono-, di- and tri-sialyloligosaccharides). Reproduced from Koketsu et al. (1993) [51] by permission of Institute of Food Technologists, Chicago.

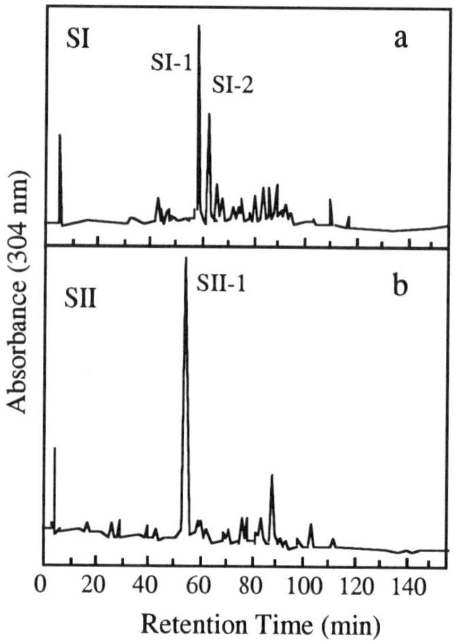

Figure 9. Separation of SI (mono-) and SII (disialyloligosacchrides) from delipidated egg yolk on a Wakosil 5C18-200 column (0.4 × 25 cm).
Eluant: NaH$_2$PO$_4$-acetonitrile (92:8, v/v); flow rate: 0.5 mL/min; detection: 304 nm. (a) and (b) show the elution profiles of SI and SII, respectively.
Reproduced from Koketsu et al. (1993) [51] by permission of Institute of Food Technologists, Chicago.

SI-1

```
               Galβ1–4GlcNAcβ1–2Manα1–6
                                         Manβ1–4GlcNAcβ1–4GlcNAc
NeuAcα2–6Galβ1–4GlcNAcβ1–2Manα1–3
```

SI-2

```
               GlcNAcβ1–2Manα1–6
                                         Manβ1–4GlcNAcβ1–4GlcNAc
NeuAcα2–6Galβ1–4GlcNAcβ1–2Manα1–3
```

SII-1

```
NeuAcα2–6Galβ1–4GlcNAcβ1–2Manα1–6
                                         Manβ1–4GlcNAcβ1–4GlcNAc
NeuAcα2–6Galβ1–4GlcNAcβ1–2Manα1–3
```

Figure 10. Structures of the major sialyloligosaccharides isolated from delipidated egg yolk. Reproduced from Koketsu et al. (1993) [51] by permission of Institute of Food Technologists, Chicago.

V. BIOLOGICAL ACTIVITIES OF EGG YOLK SIALYLOLIGOSACCHARIDES

Sialylglycoconjugates, such as gangliosides, sialyloligosaccharides, and sialylglycoproteins, have been reported to play various important roles in animal and human tissue cells. For example, they act as receptors of such viruses as Sendai virus [57], influenza virus [58-60], coronavirus [61], etc. We studied hen egg sialylglycoconjugates to apply them to the fields of functional foods and drugs [62, 63].

A. ROTAVIRAL INHIBITORY EFFECT OF EGG YOLK SIALYLOLIGOSACCHARIDES

Rotavirus is known as a major pathogen of infectious gastroenteritis in infants. It has been reported that more than one million infants are killed annually by rotaviral gastroenteritis in developing countries [64]. Also, a large number of children of the world suffer from diarrhea and vomiting caused by infection of rotavirus every year [65-68].

A study of vaccines for prevention of rotaviral infection has been made so far [69, 70]. But the vaccination has remained unsuccessful because of the difficulty of inducing specific antibody in the intestinal tracts of infants whose immunity generally has not developed yet [71]. The prevention of rotaviral infection has been long awaited.

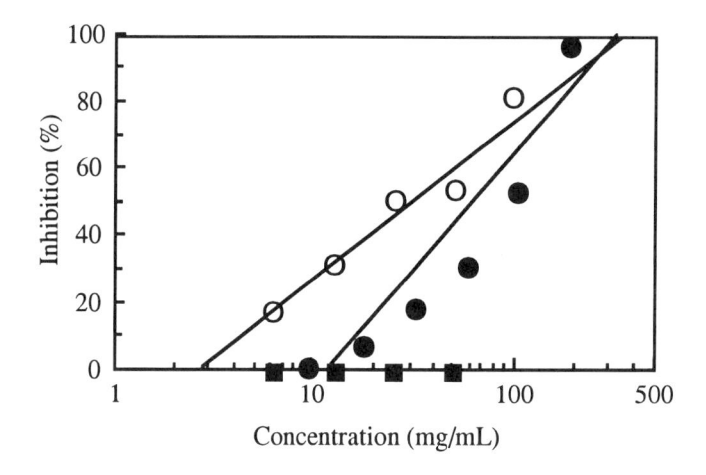

Figure 11. The inhibitory effect of sialyloligosaccharide and neutral oligosaccharide fractions on rotaviral (SA-11) infection. -O-, Sialyloligosaccharide fraction; -●-, oligosacchride fraction; -■-, neutral oligosaccharide fraction. Reproduced from Koketsu et al. (1995) [62] by permission of American Chemical Society, Washington, D. C.

A suspension of delipidated egg yolk was incubated with *Bacillus* neutral proteinase to release the peptides which bear oligosaccharides from glycoproteins. The peptidyl oligosaccharides appeared in the supernatant of the suspension, and the concentrate was found to inhibit rotaviral infection *in vitro* [72, 73]. The IC_{50} of this fraction was 60 mg/ml (Figure 11). The fraction was subjected to chromatography to separate sialyloligosaccharides (acidic oligosaccharides) and neutral oligosaccharides on an anion exchanger. The sialyloligosaccharide fraction inhibited rotaviral replication and its IC_{50} was 31.9 mg/ml, whereas the neutral oligosaccharide fraction showed no inhibition (Figure 11). The peptidyl sialyloligosaccharide fraction and sialyloligosaccharide fraction were incubated with neuraminidase, but the asialo products showed no inhibitory effect on rotaviral propagation. These results clearly indicate that the rotaviral inhibition shown by the peptide linked oligosaccharide fraction and sialyloligosaccharide fraction, respectively, is attributable to sialic acid.

The inhibitory effect of the sialyloligosaccharide fraction on the rotavirus was investigated *in vivo* using suckling mice [74] which were previously infected with rotavirus SA-11 (4.4 × 10^5 FCFU/mouse) 3 hours before administration of the sialyloligosaccharide fraction. The incidence of diarrhea was inspected 1, 3, and 5 days after oral administration of 2.5 mg sialyloligosaccharide fraction/mouse. The group administered the sialyloligosaccharide showed a significantly lower incidence of rotaviral diarrhea by 24 % and 43% on the 3rd day and 5th day, respectively, compared with the control group (Figure 12). The egg yolk sialyloligosaccharides can be prepared by the method mentioned above, and the oral administration of sialyloligosaccharides isolated from hen egg which is one of the most popular foods, may be useful for the prevention of rotaviral infection.

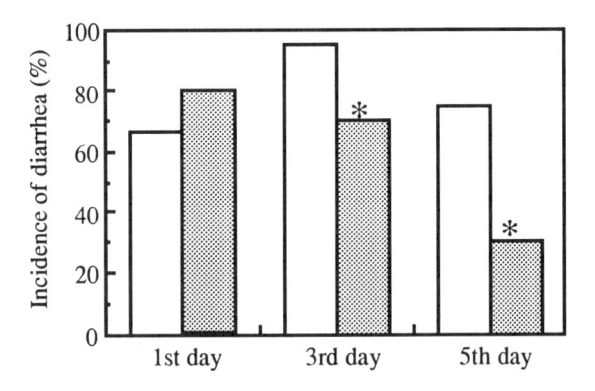

Figure 12. Effect of the administration of sialyloligosaccharide fraction to suckling mice previously inoculated with rotavirus (SA-11). Left bar: saline; right bar: sialyloligosaccharide fraction. (2.5mg dose per head). * $p< 0.05$
Reproduced from Koketsu et al. (1995) [62] by permission of American Chemical Society, Washington, D. C.

B. LEARNING PERFORMANCE OF SIALYLOLIGOSACCHARIDES

In order to confirm the effect of the administration of sialyloligosaccharides to infant rats during the lactation period, their learning performance with or without the intake of sialyloligosaccharides was examined by the maze test [63] (Figure 13). The egg yolk sialyloligosaccharides were administrated to rats age 14 days to 21 days after birth, and the goal reaching time and success ratio of goal reaching were inspected for 42-49 day old rats by applying the maze test. The result revealed that the group administered with egg yolk sialyloligosaccharide fraction was higher in the success ratio of goal reaching than the control group and also, the goal reaching time was significantly shorter than that of the control group (Figure 13). This result suggests that the sialyloligosaccharides play an important role in improving the learning performance of infant rats. However, the intake of sialyloligosaccharides should begin during the lactation period because not only is the learning performance of rats improved, but also the rats are protected from various infectious diseases during their growth.

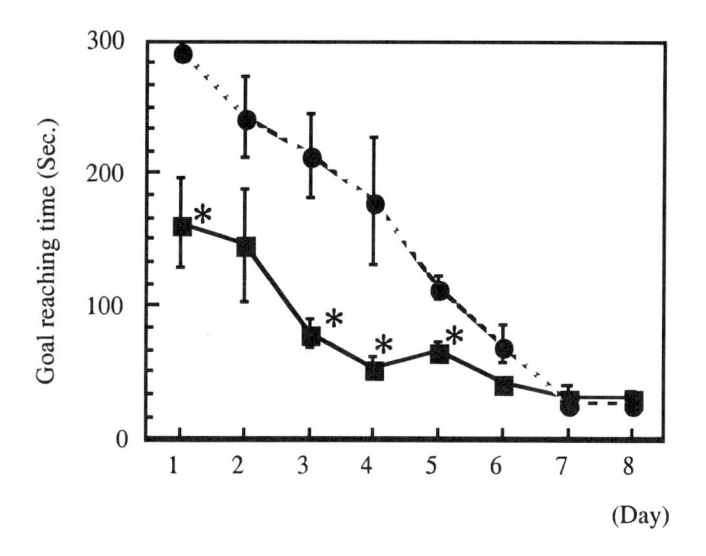

Figure 13. Effect of oral administration of sialyloligosaccharide fraction on goal reaching time of rats examined by the maze test. -■-, sialyloligosaccharide fraction; -●-, saline. Reproduced from Koketsu et al. (1995) [62] by permission of Nihon Oyotoshitu Kagaku Kai.

REFERENCES

1. **Sharon, N.**, *Complex Carbohydrates. Their Chemistry, Biosynthesis and Functions*, Addison-Wesley, New York, 1975.
2. **Li, S. C., Chien, J. L., Wan, C. C., and Li, Y. T.**, Occurence of glycosphingolipids in chicken egg yolk, *Biochem. J.*, **173**, 697, 1978.
3. **Schauer, R.**, Chemistry, metabolism, and biological functions of sialic acids, *Adv. Carbohydr. Chem. Biochem.*, **40**, 131, 1982.
4. **Warren, L.**, Destribution of sialic acids in nature, *Comp. Biochem. Physiol.*, **10**, 153, 1963.
5. **Ng, S. S. and Dain, J. A.**, The natural occurrence of sialic acids, in *Biological Roles of Sialic Acid*, Rosenber, A. and Schengrund, C. L., Eds., Plenum, New York, 1976, p 59.
6. **Kedzierska, B.**, *N*-acetylneuraminic acid: A constituent of the lipopolysaccharide of *Salmonella toucra*, *Eur. J. Biochem.*, **91**, 545, 1978.
7. **Juneja, L. R., Koketsu, M., Nishimoto, K., Kim, M., Yamamoto, T., and Itoh, T.**, Large-scale preparation of sialic acid from chalaza and egg-yolk membrane, *Carbohydr. Res.*, **214**, 179, 1991.
8. **Koketsu, M., Juneja, L. R., Kawanami, H., Kim, M., and Yamamoto, T.**, Preparation of *N*-acetylneuraminic acid from delipidated egg yolk, *Glycoconjugate J.*, **9**, 70, 1992.
9. **Everson, G. J. and Souders, H. J.**, Composition and nutritive importance of eggs, *J. Am. Diet. Assoc.*, **33**, 1244, 1957.
10. **Ayres, J. C. and Stewert, G. F.**, Removal of sugar from raw egg white by yeast before

drying, *Food Technol.*, **1**, 519, 1947.
11. **Lin, Y. and Feeney, R. E.**, *Glycoproteins*, Elsevier/North-Holland Biomedical Press, Amsterdam, 1972.
12. **Beeley, J. G.**, Location of the carbohydrate groups of ovomucid, *Biochem. J.*, **159**, 335, 1976.
13. **Kato, I., Schrode, J., Kohr, W. J., and Laskowski, M. J.**, Chicken ovomucoid: Determination of its amino acid sequence, determination of the trypsin reactive site, and preparation of all three of its domains, *Biochemistry*, **26**, 193, 1987.
14. **Yamashita, K., Kamerling, J. P., and Kobata, A.**, Structural study of the carbohydrate moiety of hen ovomucoid. Occurrence of a series of pentaantennary complex-type asparagine-linked sugar chains, *J. Biol. Chem.*, **257**, 12809, 1982.
15. **Conchie, J., Hay, A. J., and Lomax, J. A.**, The carbohydrate units of asialo-ovomucoid: Structural features, *Carbohydr. Res.*, **112**, 281, 1983.
16. **Conchie, J. and Hay, A. J.**, The carbohydrate units of asialo-ovomucoid: Their heterogeneity and enzymic degradation, *Carbohydr. Res.*, **112**, 261, 1983.
17. **Yamashita, K., Kamerling, J. P., and Kobata, A.**, Structural studies of the sugar chains of hen ovomucoid. Evidence indicating they are formed mainly by the alternate biosynthetic pathway of asparagine-linked sugar chains, *J. Biol. Chem.*, **258**, 2099, 1983.
18. **Parente, J. P., Leroy, Y., Montreuil, J., and Fournet, B.**, Separation of sialyl-oligosaccharides by high-performance liquid chromatogtaphy. Application to analysis of carbohydrate units of acidic oligosaccharides obtained by hydrazinolysis of hen ovomucoid, *J. Chromatogr.*, **288**, 147, 1984.
19. **Yamashita, K., Tachibana, Y., and Hitoi, A.**, Sialic acid-containing sugar chains of hen ovalbumin and ovomucoid, *Carbohydr. Res.*, **130**, 271, 1984.
20. **Yet, M.-G., Chin, C. C. Q., and Wold, F.**, The covalent structure of individual N-linked glycopeptides from ovomucoid and asialofetuin, *J. Biol. Chem.*, **263**, 111, 1988.
21. **McReynolds, L., O'Malley, B. W., Nisbet, A. D., Fothergill, J. E., Givol, D., Fields, S., Robertson, M., and Brownlee, G. G.**, Sequence of chicken ovalbumin mRNA, *Nature*, **273**, 723, 1978.
22. **Nisbet, A. D., Saundry, R. H., Moir, A. J. G., Fothergill, L. A., and Fothergill, J. E.**, The complete amino acid sequence of hen ovalbumin, *Eur. J. Biochem.*, **115**, 335, 1981.
23. **Tai, T., Yamashita, K., Ogata-Arakawa, M., Koide, N., Muramatsu, T., Iwashita, S., Inoue, Y., and Kobata, A.**, Structural studies of two ovalbumin glycopeptides in relation to the endo-β-N-acetylglucosaminidase specificity, *J. Biol. Chem.*, **250**, 8569, 1975.
24. **Tai, T., Yamashita, K., Ito, S., and Kobata, A.**, Structures of the carbohydrate moiety of ovalbumin glycopeptide III and the difference in specificity of endo-β-N-acetylglucosaminidases CII and H, *J. Biol. Chem.*, **252**, 6687, 1977.
25. **Tai, T., Yamashita, K., and Kobata, A.**, The substrate specificities of endo-β-N-acetylglucosaminidases CII and H, *Biochem. Biophys. Res. Commun.*, **78**, 434, 1977.
26. **Yamashita, K., Tachibana, Y., and Kobata, A.**, The structures of the galactose-containing sugar chains of ovalbumin, *J. Biol. Chem.*, **253**, 3862, 1978.
27. **Wang, T. H., Chen, T. F., and Barofsky, D. F.**, Mass spectrometry of L-β-aspartamido carbohydrates isolated from ovalbumin, *Biomed. Environmental Mass Spectrom.*, **16**, 335, 1988.
28. **Honda, S., Makino, A., Suzuki, S., and Kakehi, K.**, Analysis of the oligosaccharides in ovalbumin by high-performance capillary electrophoresis, *Anal. Biochem.*, **191**, 228, 1990.
29. **Williams, J., Elleman, T. C., Kingston, I. B., Wilkins, A. G., and Kuhn, K. A.**, The primary structure of hen ovotransferrin, *Eur. J. Biochem.*, **122**, 297, 1982.
30. **Dorland, L., Haverkamp, J., Vliegenthart, J. F. G., Spik, G., Fournet, B., and**

Montreuil, J., Investigation by 360-MHz ^{1}H-nuclear-magnetic-resonance spectroscopy and methylation analysis of the single glycan chain of chicken ovotransferrin, *Eur. J. Biochem.*, **100**, 569, 1979.
31. **Spik, G., Coddeville, B., and Montreuil, J.,** Comparative study of the primary structures of sero-, lacto- and ovotransferrin glycans from different species, *Biochimie*, **70**, 1459, 1988.
32. **Clark, R. C.,** The isolation and composition of two phosphoproteins from hen's egg, *Biochem. J.*, **118**, 537, 1970.
33. **Wallace, R. A. and Morgan, J. P.,** Chromatographic resolution of chicken phosvitin. Multiple macromolecular speicies in a classic vitellogenin-derived phosphoprotein, *Biochem. J.*, **240**, 871, 1986.
34. **Brockbank, R. L. and Vogel, H. J.,** Structure of the oligosaccharide of hen phosvitin as determined by two-dimensional ^{1}H NMR of the intact glycoprotein, *Biochemistry*, **29**, 5574, 1990.
35. **Shainkin, R. and Perlmann, G. E.,** Phosvitin, a phosphoglycoprotein: Composition and partial structure of carbohydrate moiety, *Arch. Biochem. Biophys.*, **145**, 693, 1971.
36. **Shainkin, R. and Perlmann, G. E.,** Phosvitin, a phosphoglycoprotein. I. Isolation and characterization of a glycopeptide from phosvitin, *J. Biol. Chem.*, **246**, 2278, 1971.
37. **Robinson, D. S. and Monsey, J. B.,** The composition and proposed subunit structure of egg-white β-ovomucin, *Biochem. J.*, **147**, 55, 1975.
38. **Hayakawa, S. and Sato, Y.,** Subunit structures of sonicated α- and β- ovomucin and their molecular weights estimated by sedimentation equilibrium, *Agric. Biol. Chem.*, **42**, 957, 1978.
39. **Kato, A., Fujinaga, K., and Yagishita, K.,** Nature of the carbohydrate side chains and their linkage to the protein in chicken egg white ovomucin, *Agric. Biol. Chem.*, **37**, 2479, 1973.
40. **Kato, A., Susumu, H., and Kobayashi, K.,** Structure of the sulfated oligosaccharide chain of ovomucin, *Agric. Biol. Chem.*, **42**, 1025, 1978.
41. **Kato, A., Susumu, H., Sato, H., and Kobayashi, K.,** Fractionation and characterization of the sulfated oligosaccharide chains of ovomucin, *Agric. Biol. Chem.*, **42**, 835, 1978.
42. **Kato, A., Miyoshi, Y., Suga, M., and Kobayashi, K.,** Separation and characterization of sulfated glycopeptides from ovomucin, chalaza and yolk membrane in chicken eggs, *Agric. Biol. Chem.*, **46**, 1285, 1982.
43. **Strecker, G., Wieruszeski, J.-M., Martel, C., and Montreuil, J.,** Complete ^{1}H- and ^{13}C-NMR assignments for two sulphated oligosaccharide alditols of hen ovomucin, *Carbohydr. Res.*, **185**, 1, 1989.
44. **Ostrowski, W., Skarzynski, B., and Zak, Z.,** Isolation and properties of flavoprotein from the egg yolk, *Biochim. Biophys. Acta*, **59**, 515, 1962.
45. **Ostrowski, W., Zak, Z., and Krawczyk, A.,** Riboflavin flavoprotein from egg yolk, *Acta Biochem. Pol.*, **15**, 241, 1968.
46. **Miller, M. S., Buss, E. G., and Clagett, C. O.,** The role of oligosaccharide in transport of egg yolk riboflavin-binding protein to the egg, *Biochim. Biophys. Acta*, **677**, 225, 1981.
47. **Hamazume, Y., Mega, T., and Ikenaka, T.,** Characterization of hen egg white- and yolk-roboflavin binding proteins and amino acid sequence of egg white-riboflavin binding protein, *J. Biochem.*, **95**, 1633, 1984.
48. **Norioka, N., Okada, T., Hamazume, Y., Mega, T., and Ikenaka, T.,** Comparison of the amino acid sequences of hen plasma-, yolk-, and white-riboflavin binding proteins, *J. Biochem.*, **97**, 19, 1985.
49. **Tatutani, M., Norioka, N., Mega, T., Hase, S., and Ikenaka, T.,** Structures of sugar

chains of hen egg yolk riboflavin-binding protein, *J. Biochem.*, **113**, 677, 1993.
50. Ohta, M., Hamako, J., Yamamoto, S., Hatta, H., Kim, M., Yamamoto, T., Oka, S., Mizuochi, T., and Matsuura, F., Structures of asparagine-linked oligosaccharides from hen egg-yolk antibody (IgY). Occurrence of unusual glucosylated oligo mannose type oligosaccharides in a mature glycoprotein, *Glycoconjugate J.*, **8**, 400, 1991.
51. Koketsu, M., Juneja, L. R., Kim, M., Ohta, M., Matsuura, F., and Yamamoto, T., Sialyloligosaccharides of egg yolk fraction, *J. Food Sci.*, **58**, 743, 1993.
52. Koketsu, M., Seko, A., Juneja, L. R., Kim, M., Kashimura, N., and Yamamoto, T., An efficient preparation and structural characterization of sialylglycopeptides from protease treated egg yolk, *J. Carbohydr. Chem.*, **9**, 833, 1995.
53. Matsuura, F. and Imaoka, A., Chromatographic separation of asparagine-linked oligosaccharides labeled with an ultraviolet absorbing compound, *p*-aminobenzoic acid ethyl ester, *Glycoconjugate J.*, **5**, 13, 1988.
54. Ohta, M., Kobatake, M., Matsumura, A., and Matsuura, F., Separation of Asn-linked sialyloligosaccharides labeled with *p*-aminobenzoic acid ethyl ester by high performance liquid chromatography, *Agric. Biol. Chem.*, **54**, 1045, 1990.
55. Matsuura, F., Ohta, M., Murakami, K., Hirano, K., and Sweeley, C. C., The combination of normal phase with reversed phase high performance liquid chromatography for the analysis of asparagine-linked neutral oligosaccharides labelled with *p*-aminobenzoic acid ethyl ester, *Biomed. Chromatogr.*, **6**, 77, 1992.
56. Vliegenthart, J. F. G., Dorland, L., and Van Halbeek, H., High-resolution, ^{1}H-nuclear magnetic resonance spectroscopy as a tool in the structural analysis of carbohydrates related to glycoproteins, *Adv. Carbohydr. Chem. Biochem.*, **41**, 209, 1983.
57. Holmgren, J., Svennerholm, L., Elwing, H., Fredman, P., and Strannegård, Ö., Sendai virus receptor: Proposed recognition structure based on binding to plastic-adsorbed gangliosides, *Proc. Natl. Acad. Sci. USA*, **77**, 1947, 1980.
58. Suzuki, Y., Nakao, T., Ito, T., Watanabe, N., Toda, Y., Guiyun, X., Suzuki, T., Kobayashi, T., Kimura, Y., Yamada, A., Sugawara, K., Nishimura, H., Kitame, F., Nakamura, K., Deya, E., Kiso, M., and Hasegawa, A., Structural determination of gangliosides that bind to influenza A, B, and C viruses by an improved binding assay: Strain-specific receptor epitopes in sialo-sugar chains, *Virology*, **189**, 121, 1992.
59. Von Itzstein, M., Wu, W. Y., Kok, G. B., Pegg, M. S., Dyason, J. C., Jin, B., Phan, T. V., Smythe, M. L., White, H. F., Oliver, S. W., Colman, P. M., Varghese, J. N., Ryan, D. M., Woods, J. M., Bethell, R. C., Hotham, V. J., Cameron, J. M., and Penn, C. R., Rational design of potent sialidase-based inhibitors of influenza virus replication, *Nature*, **363**, 418, 1993.
60. Zimmer, G., Suguri, T., Reuter, G., Yu, R. K., Schauer, R., and Herrler, G., Modification of sialic acids by 9-*O*-acetylation is detected in human leucocytes using the lectin property of influenza C virus, *Glycobiology*, **4**, 343, 1994.
61. Schultze, B. and Herrler, G., Recognition of cellular receptors by bovine coronavirus, *Arch. Virol.*, **9**, 451, 1994.
62. Koketsu, M., Nitoda, T., Juneja, L. R., Kim, M., Kashimura, N., and Yamamoto, T., Sialylglycopeptides from egg yolk as an inhibitor of rotaviral infection, *J. Agric. Food Chem.*, **43**, 858, 1995.
63. Koketsu, M., Nakata, K., Juneja, L. R., Kim, M., and Yamamoto, T., Learning performance of egg yolk sialyloligosaccharides, *Oyotoshitsu Kagaku (in Japanese)*, **9**, 15, 1995.
64. Kaspikian, A. Z., Flores, J., Vesikari, T., Ruuska, T., Madore, H. P., Green, K. Y., Gorziglia, M., Hoshino, Y., Chanock, R. M., Midthun, K., and Perez-Schael, I., Recent

advances in development of a rotavirus vaccine for prevention of severe diarrheal illness of infants and young children, in *Advances in Experimental Medical Biology*, **Vol. 310**, Whistler, R. L. and Wolfrom, M. L., Eds., Academic Press, New York, 1991, p 255.
65. **Gouvea, V., Decastro, L., Timenetsky, M. D., Greenberg, H., and Santos, N.**, Rotavirus serotype G5 associated with diarrhea in Brazilian children, *J. Clin. Microbiol.*, **32**, 1408, 1994.
66. **Kaga, E., Iizuka, M., Nakagomi, T., and Nakagomi, O.**, The distribution of G (Vp7) and P (Vp4) serotype among human rotaviruses recovered from Japanese children with diarrhea, *Microbiol. Immunol.*, **38**, 317, 1994.
67. **Noel, J., Mansoor, A., Thaker, U., Hermann, J., Perronhenry, D., and Cubitt, W. D.**, Identification of adenoviruses in feces from patients with diarrhea at the hospitals for sick children, London, 1989-1992, *J. Med. Virol.*, **43**, 84, 1994.
68. **Tabassum, S., Shears, P., and Hart, C. A.**, Genomic characterization of rotavirus strains obtained from hospitalized children with diarrhea in Bangladesh, *J. Med. Virol.*, **43**, 50, 1994.
69. **Greenberg, H. B.**, Rotavirus vaccination-Current status- A brief summary, in *Biotechnology R&D Trends*, **Vol. 700**, Tzotzos, G. T., Ed., New York Academy of Sciences, New York, 1993, p 32.
70. **Forrest, B. D.**, Diarrhoeal disease and vaccine development, *Trans R. Soc. Trop. Med. Hyg.*, **87**, 39, 1993.
71. **DeMol, P., Zissis, G., Butzler, J. P., Mutwewingabo, A., and André, F. E.**, Failure of live, attenuated oral rotavirus vaccine, *Lancet*, **2**, 108, 1986.
72. **Fukudome, K., Yoshie, O. and Konno, T.**, Comparison of human, simian, and bovine rotaviruses for requirement of sialic acid in hemagglutination and cell adsorption, *Virology*, **172**, 196, 1989.
73. **Svensson, L.**, Group C rotavirus requires sialic acid for erythrocyte and cell receptor binding, *J. Virol.*, **66**, 5582, 1992.
74. **Hatta, H., Tsuda, K., Akachi, S., Kim, M., Yamamoto, T., and Ebina, T.**, Oral passive immunization effect of antihuman rotavirus IgY and its behavior against proteolytic enzymes, *Biosci. Biotech. Biochem.*, **57**, 10770, 1993.

Chapter 8

CHEMICAL AND PHYSICOCHEMICAL PROPERTIES OF HEN EGGS AND THEIR APPLICATION IN FOODS

H. Hatta, T. Hagi, and K. Hirano

TABLE OF CONTENTS

I. Introduction
II. Heat Induced Gelation
    A. Gelation and Agglutination
    B. Factors Affecting Heat Induced Gelation of Egg Albumen
        1. Evaluation Method of Heat Induced Gelation
        2. Effects of Protein Concentration and Temperature
        3. Effects of pH
        4. Effects of Addition of Salt and Sucrose
    C. Heat Induced Gelation of Ovalbumin
        1. Formation of Transparent and Turbid Gels
        2. Mechanism of Formation of Ovalbumin Gel by Heating
        3. Structural Changes of Ovalbumin Molecule on Heating
III. Foaming Property
    A. Formation of Foam and Its Stability
    B. Foaming Ability of Egg Albumen
        1. Method for Evaluation of Foaming Ability
        2. Influence of Freshness of Egg Albumen
        3. Effect of Pasteurization
        4. Other Factors Affecting Foaming of Egg Albumen
    C. Foaming Ability of Egg White Proteins
    D. Mechanism of Foaming of Ovalbumin
IV. Emulsification
    A. Emulsification and Emulsified Foods
    B. Characteristics of Egg Yolk as an Emulsifier
        1. Method of Evaluation of Emulsifying Properties
        2. Effects of Freezing of Egg Yolk
        3. Effect of Pasteurization
        4. Effect of Acids or Salts
        5. Effects of Drying Egg Yolk
    C. Emulsifying Properties of Egg Yolk Lipoproteins

References

I. INTRODUCTION

Since ancient times, eggs have been consumed as food or utilized as ingredient in various food products. People's diets have been nutritionally enriched by learning the cooking or processing eggs for food products in various ways. Egg shows magnificent properties on cooking. These are the reasons why eggs are widely utilized in cooking and manufacturing of various processed foods or confectioneries.

In the food industry, the important functional properties of eggs are: the heat induced gelation and foaming properties of egg albumen and the emulsifying properties of egg yolk. Heating, foaming, or emulsifying treatment of eggs cause changes in the structure of egg white proteins and egg yolk lipoproteins which result in changes in functional properties such as capacity for holding water, air, or oil (Figure 1). In Japan, heat induced gelation of egg albumen has been applied for processing ham, sausages, boiled fish paste (kamaboko), and noodles. The foaming property of egg is used in making cakes and meringue, and the emulsifying property for mayonnaise and salad dressings, etc.

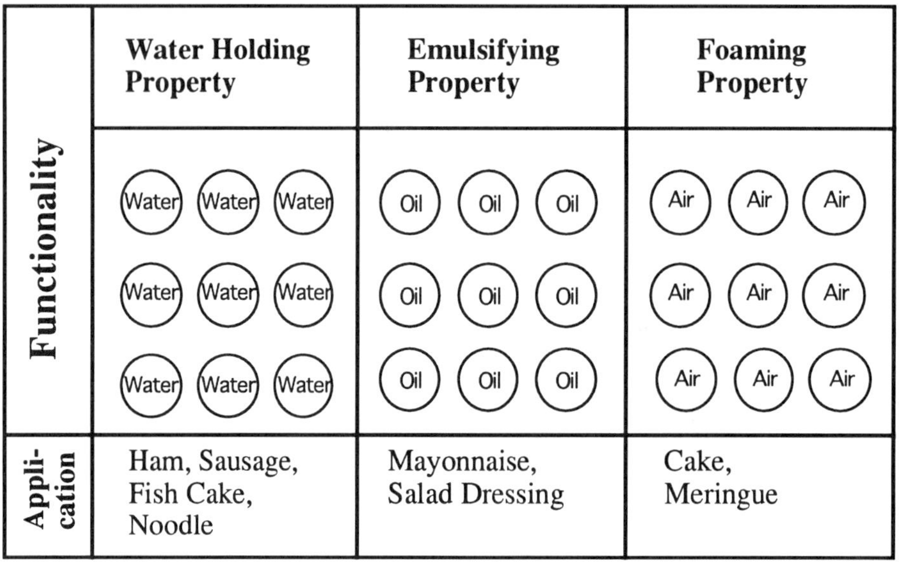

	Water Holding Property	**Emulsifying Property**	**Foaming Property**
Functionality	(Water)(Water)(Water) (Water)(Water)(Water) (Water)(Water)(Water)	(Oil)(Oil)(Oil) (Oil)(Oil)(Oil) (Oil)(Oil)(Oil)	(Air)(Air)(Air) (Air)(Air)(Air) (Air)(Air)(Air)
Application	Ham, Sausage, Fish Cake, Noodle	Mayonnaise, Salad Dressing	Cake, Meringue

Figure 1. Schematic representation of characteristic properties of egg proteins applied to cooking.

In this chapter, the research work carried out on the influence of salts, pH, and storage time on the functional properties of hen eggs, which are important for the food industry, are described. Some recent information about the relationship between the protein structures of eggs and their functional properties is also discussed.

II. HEAT INDUCED GELATION

A. GELATION AND AGGLUTINATION

Egg albumen is a colloidal mixture of various proteins. Some physical and chemical factors (temperature, pH, ionic strength, protein concentration, etc.) affect the structure of proteins causing agglutination or gelation. Agglutination occurs by strong binding forces working among protein molecules, followed by drainage of water, and finally resulting in the formation of giant precipitable molecules. On the other hand, gelation occurs when a balance is established between the intermolecular binding force of colloidal molecules and the solvent binding force of the colloid particles. In the case of gelation of protein, the protein molecules bind to each other at certain sites to make a jungle gym-like structure, the spaces which are filled with solvent linked by hydrogen bonds to the protein molecules.

B. FACTORS AFFECTING HEAT INDUCED GELATION OF EGG ALBUMEN
1. Evaluation method of heat induced gelation

In the food industry, gel strength and water holding capacity are the parameters for evaluation of heat induced gelation. Gel strength is generally evaluated by measuring the mechanical force necessary to break the gel surface of a definite area (g/cm^2). The water holding capacity is determined by weighing the solvent transferred to a filter paper when a cube of the gel is placed on the filter paper under a fixed condition.

2. Effects of protein concentration and temperature

It has been known that the denaturation temperatures of several egg white proteins are 84°C for ovalbumin, 61°C for ovotransferrin, 77°C for ovomucoid, and 75°C for lysozyme [1]. The gelation of egg albumen, which is the result on average of diverse denaturation temperatures of various proteins, begins at 60-65°C but, at this temperature, the gel formed is too soft to be stabilized even if heated for a long time. At 70°C, gelation proceeds rapidly and the gel formed is hard.

Holt and his co-workers investigated the texture of egg albumen gels produced by heating at several temperatures from 65 to 90°C in a neutral pH region. They found that gel strength and water holding capacity decreased at temperatures over 80°C. They suggested that temperatures higher than 80°C should not be employed for obtaining a favorable texture of egg albumen gel [2]. Ovalbumin, the major protein of egg albumen, is denatured at 84°C, thus heating at 90°C brings an exhaustive denaturation resulting in intensive protein agglutination. As the protein concentration of egg white decreases, higher temperature is necessary for gelation to a certain extent, but the texture of the gel formed is soft. On the other hand, higher protein concentrations of egg albumen bring gelation at lower temperatures, and its texture is relatively hard.

To evaluate gel strength, egg white powder was dissolved into 2% NaCl solution in order to get protein concentrations of 7.5 %, 10.0 %, 12.5 %, or 15.0 %. These solutions were heated at 75°C for 60 min. The strengths of the gels produced were 222, 436, 760, and 1,062 g/cm^2, respectively. This is because at higher protein concentrations, the opportunities for the chemical and physicochemical binding forces to work among protein molecules increase, generating gelation with more dense spaces in the jungle gym-like structure. Electron microscopic observation revealed that egg albumen gels of higher gel strengths are constructed with more dense network structures (Figure 2).

3. Effects of pH

The pH value of very fresh egg albumen is generally 7.6-7.9. Upon storing eggs at 25°C for 6 days, the pH rises to 9.2-9.5. During storage, carbon dioxide is released from the eggs, resulting in rising pH values of the albumen. Ogawa and Tanabe have reported on the relation between the freshness of eggs and heat induced egg albumen gel that a fall in the freshness of eggs (meaning a rise in pH value of egg albumen) is accompanied by a decrease of egg albumen gel strength [3].

We prepared several gels using 15% (w/w) egg albumen powder dissolved at several pHs and then heated at 90°C for 30 min. The gels obtained were tested for gel strength and the percentage of water released from the gels (Figure 3). The gel obtained at pH 5 showed the highest gel strength and also the highest percentage of water release. As the pH increased

A   B   C

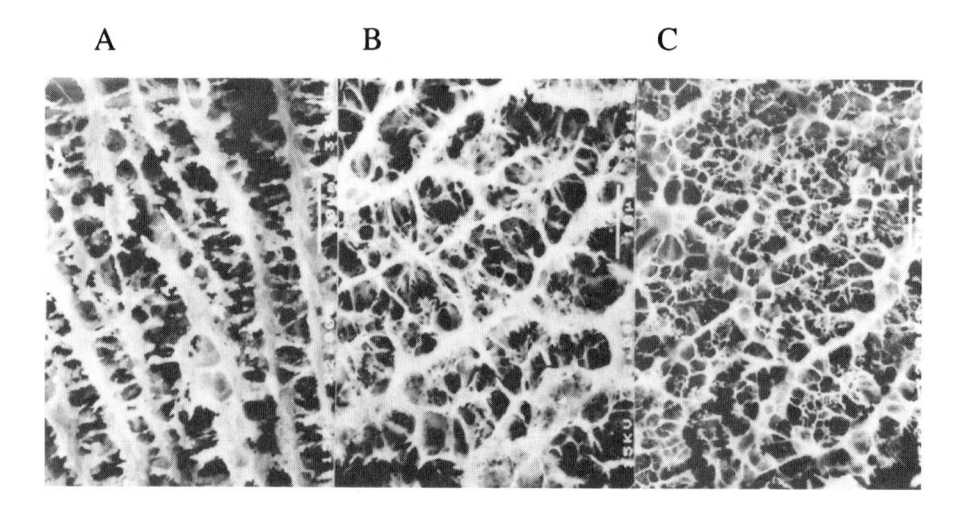

Figure 2. Scaning electron micrographs of cryofractured surfaces of heat induced egg albumen gels. Three kinds of egg albumen powder (A, B, and C preparations) were dissolved at 15% (w/w) into 2% NaCl solution (pH 7.0), respectively. The albumen solutions were heated at 75°C for 1 hr, then cooled at 20°C for 1 hr.

Gel strength (g/cm^2) : (A) 250, (B) 1,100, and (C) 1,800; magnification: x 2,000.

from 5 to 10, the gel strength slightly decreased and the degree of water release decreased. This fact indicates that the water holding capacity increased with an increase in pH value. At the pH value of the isoelectric point (pI) of a protein, the number of positive and negative electric charges of a protein are equal. Therefore, the electrostatic repulsive forces are minimum. This situation facilitates agglutination of the protein. The isoelectric points of the proteins constituting egg albumen have been reported to be as follows: ovalbumin, 4.5; ovotransferrin, 6.05; ovomucoid, 4.1; and ovomucin, 4.5-5.0; This shows that their isoelectric points are slightly on the acidic side.

In the investigation of the effects of variation of pH and temperature on egg albumen gelation, Sato and his co-workers reported that the lowest temperature for gelation is obtained at pH 5.5, and, at pH values over 9.0, the temperature for gelation becomes higher [4].

4. Effects of addition of salt and sucrose

Egg albumen gels were prepared using 15% egg albumen powder suspension with the addition of various concentrations of NaCl (pH 7.0) and heating them at 90°C for 30 min. The gel strength and the percentage of water release observed are shown in Figure 4. The strongest gel was obtained at 0.5% NaCl. The higher the concentration of NaCl, the less gel strength resulted. On the other hand, the release of water was not affected by the concentration of NaCl.

Since proteins are ampholyte substances, their electrostatic charges are influenced by the addition of electrolytes. The presence of salts such as NaCl suppress the net charge of the protein molecules, resulting in a decrease in the electrostatic repulsive force working among the protein molecules. Therefore, the hydrophobic interaction generated among heat denatured protein molecules is relatively strengthened by the presence of salt. Therefore, the addition of salts in an appropriate amounts increases the gel formation capacity of egg albumen. However,

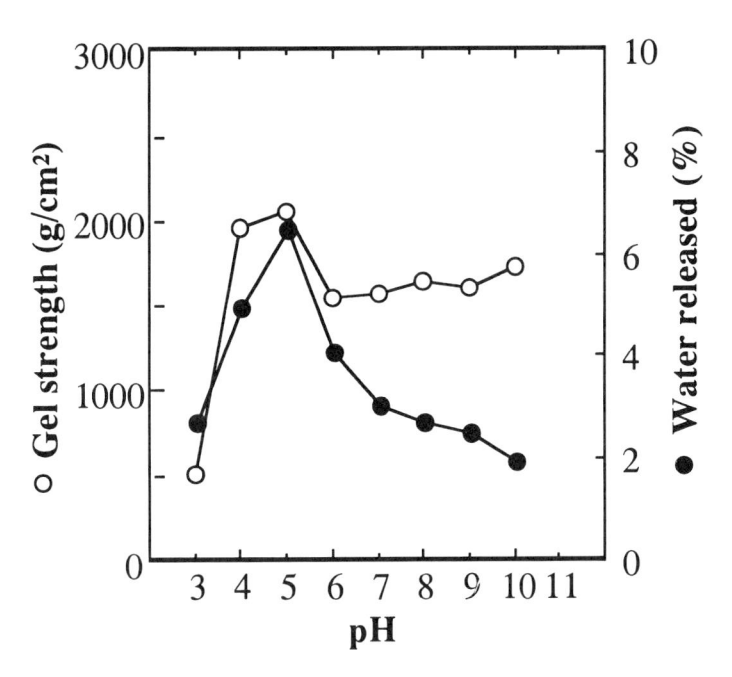

Figure 3. Properties of egg albumen gels prepared at different pHs.
Solutions of 15% (w/w) egg albumen powder at various pHs were heated at 90°C for 30 min, and, after cooling at 20°C for 1 h, their gel strength and water holding capacities were determined.

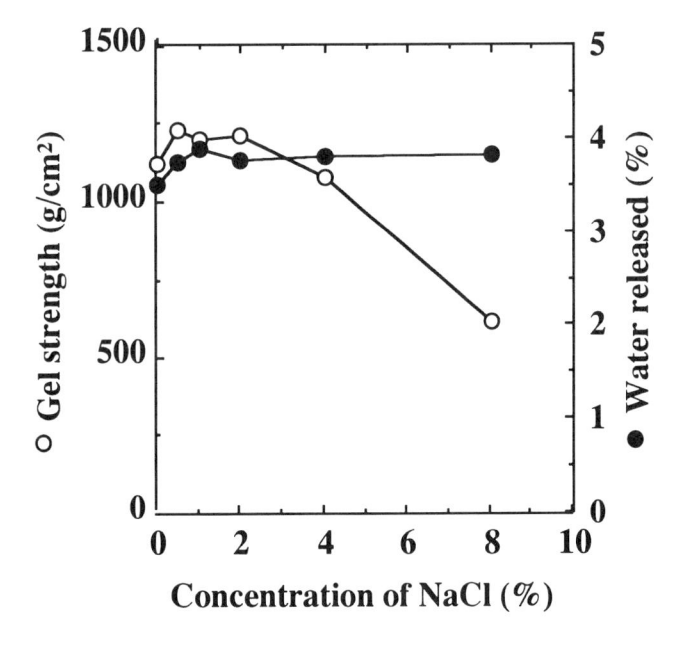

Figure 4. Properties of egg albumen gels prepared in the presence of various concentrations of NaCl.
Solutions of 15% (w/w) egg albumen powder added to various concentrations of NaCl, were heated at 90°C for 30 min, and, after cooling at 20°C for 1 h, their gel strengths and water holding capacities were determined.

the addition of excessive amounts of salt produces the inverse effect.

In general, heat denaturation of protein is considerably suppressed by the addition of sugars like sucrose or sorbitol. Since sugars have strong affinity with water, they influence the hydrophobic interaction between protein molecules. Therefore, the addition of sucrose raises the gelation temperature of egg albumen [5]. Moreover, the addition of sucrose produces soft gels in an evaluation system using baked custard [6].

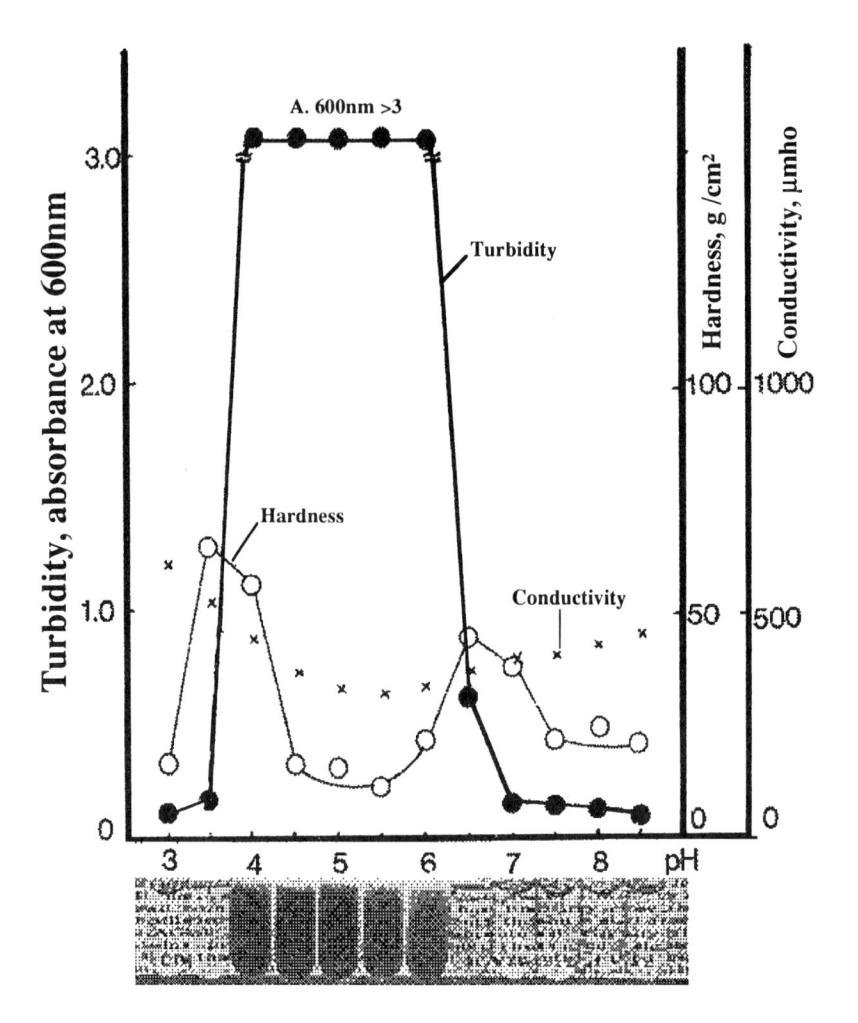

Figure 5. Hardness and turbidity of heat induced gels from ovalbumin as a function of pH value.
A solution of ovalbumin, dialyzed with deionized water, with 20 mM NaCl added, and, after adjusting the pH at various values, (final conc. of ovalbumin, 5%) was heated at 80°C for 1 h for gelation. Hatta et al. (1986) [7], reproduced by permission.

C. HEAT INDUCED GELATION OF OVALBUMIN
1. Formation of transparent and turbid gels

Ovalbumin (OVA) is a globular protein with a molecular weight of 45.5 kDa. It is a single strand polypeptide containing a small amount of phosphoric acid and carbohydrates. Since ovalbumin makes up 54% of egg white protein, its behavior dominantly affects the gelation of egg albumen. We investigated the properties of heat induced gels of ovalbumin under various conditions [7, 8]. A sufficiently dialyzed ovalbumin solution (salt-free OVA) was adjusted to pH 7.0 with NaOH and heated at 80°C for 1 h. No gelation was observed, however, the viscosity of the solution slightly increased but keeping its transparency. The pH of the salt-free OVA solution was varied and heated at 80°C for 1 h. The gels produced at pH 4.5-5.5 became turbid and the gel strength was very weak. The turbidity decreased drastically at pH values between 3.0 to 4.0 and at pH values between 6.0 to 7.0. Opaque gels were obtained at pH 4.0 and pH 6.0, and transparent gels were obtained at pH 3.5, 3.0, and at pH 6.5 and more alkaline pH values. The peaks of gel strength appeared at pH 3.5-4.0 and at pH 6.5-7.0 where the gels were transparent (Figure 5).

The gelation of OVA by heating was also examined as a function of ionic strength. The OVA solution remained as a transparent solution when heated (80°C for 1 h) at pH 3.5 with no addition of NaCl. However, when heated in the presence of 10-30 mM NaCl at pH 3.5, a transparent gel was produced. In the presence of 30-80 mM NaCl, the gel produced was turbid. On the other hand, at pH 5.5 with no addition of NaCl, a turbid agglutination occurred followed by sedimentation of protein clots. In the presence of 10-80 mM NaCl at pH 5.5, turbid gels with very low gel strength and water holding capacity were produced. At pH 7.5, in the presence of 0-10 mM NaCl, the OVA solution remained as a transparent solution. In the presence of 20-50 mM and 60-80 mM NaCl, a transparent and a turbid gel were obtained, respectively. Here again, the gel of the highest gel strength was obtained in a transparent or opaque state (Figure 6). The resolubilization of gels or clots of OVA induced by heating under various conditions was also investigated. The gels or clots produced under the above conditions were found soluble in a solution containing 1% SDS. However, they were not soluble in solutions containing urea or 2-mercaptoethanol (Figure 7). This result may be an indication that the formation of heat induced gels or clots of OVA is mainly due to elaboration of hydrophobic affinity that acts between the heat denatured OVA molecules.

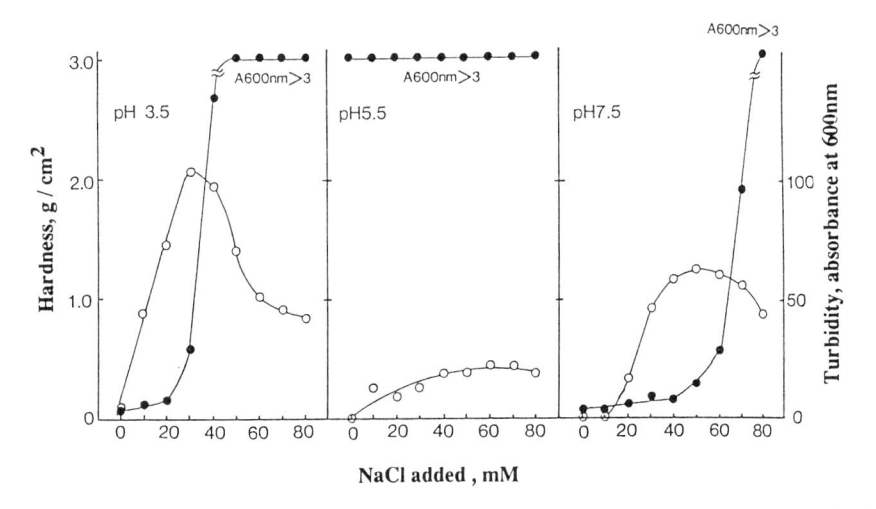

Figure 6. Hardness and turbidity of heat induced ovalbumin gels produced in the presence of various NaCl concentrations. Dialyzed ovalbumin solutions (5%), the pHs adjusted as indicated, respectively, were heated at 80°C for 1 h in the presence of various concentrations of NaCl. Reproduced from Hatta et al. (1986) [7] by permission of Japan Society for Bioscience, Biotechnology, and Agrochemistry, Tokyo.

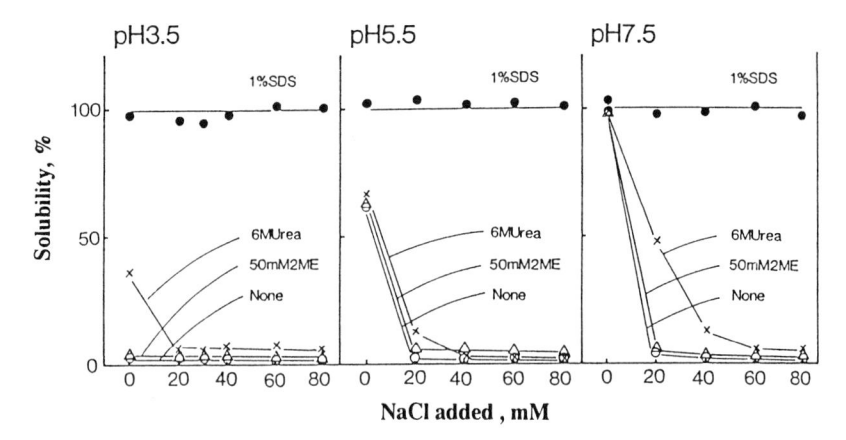

Figure 7. Solubilities of ovalbumin gels prepared by heating under different conditions in regard to pH and NaCl concentration. A 0.2 ml sample of each of the gels induced by heating at 80°C for 1 hr under above conditions was added to 1.8 ml buffer solution and heated at 37°C for 15 hr on a shaker. The mixture was then centrifuged at 3,000 g for 20 min and the absorbance of the supernatant was determined at 280 nm.

(○): 0.1M sodium phosphate, pH 7.2 (buffer: A)
(●): buffer A containing 1% SDS
(△): buffer A containing 50 mM 2-mercaptoethanol
(x): buffer A containing 6 M urea

Reproduced from Hatta et al. (1986) [7] by permission of Japan Society for Bioscience, Biotechnology, and Agrochemistry, Tokyo.

2. Mechanism of formation of ovalbumin gel by heating

The heat denaturation of OVA is initiated by breaking noncovalent bonds, in particular, intramolecular hydrogen bonds. The hydrophobic regions packed inside the native OVA are exposed to the surface of the molecules, thus, generating intermolecular hydrophobic interaction of heat denatured OVA. The conditions of the OVA solution (pH or ion strength) affect the electrostatic charge on the surface of the OVA molecules. For example, at pH values away from the isoelectric point of OVA, electrostatic repulsive forces (ERF) due to cationic ($-NH_3^+$, at acid sites) or anionic groups ($-COO^-$, at alkaline sites), become much stronger and overwhelm the intermolecular hydrophobic interaction (IHI) of the heat denatured OVA. This generates soluble OVA polymers without gelation (Figure 8). Using electronic microscopy, Koseki and his co-workers observed soluble OVA polymers like a string of beads which were obtained by heating salt-free OVA at pH values away from the isoelectric point of the OVA [9]. On the other hand, in a vicinity around the isoelectric point, the net charge of OVA molecules is zero or close to zero, so that the intermolecular hydrophobic interaction becomes much stronger than the electrostatic repulsive forces of heat denatured OVA. Therefore, the heat denatured OVA become tangled strongly to produce clots or agglutination.

As shown in Figure 8, gelation of OVA occurs at pH regions not so far from the isoelectric point and at appropriate concentrations of NaCl, where IHI and ERF were well balanced. Transparent gels are obtained when IHI is slightly lower than ERF. At this stage, the hardness and water holding capacity of the gels are at their highest, because of the fine structure of the network of the gels. On the other hand, when IHI is equal or slightly higher than ERF, opaque and turbid gels are produced. Under this condition, the denatured OVA molecules can form some clots in the structure of the gel network. Therefore, the hardness and water holding capacity are not as strong as in transparent gels.

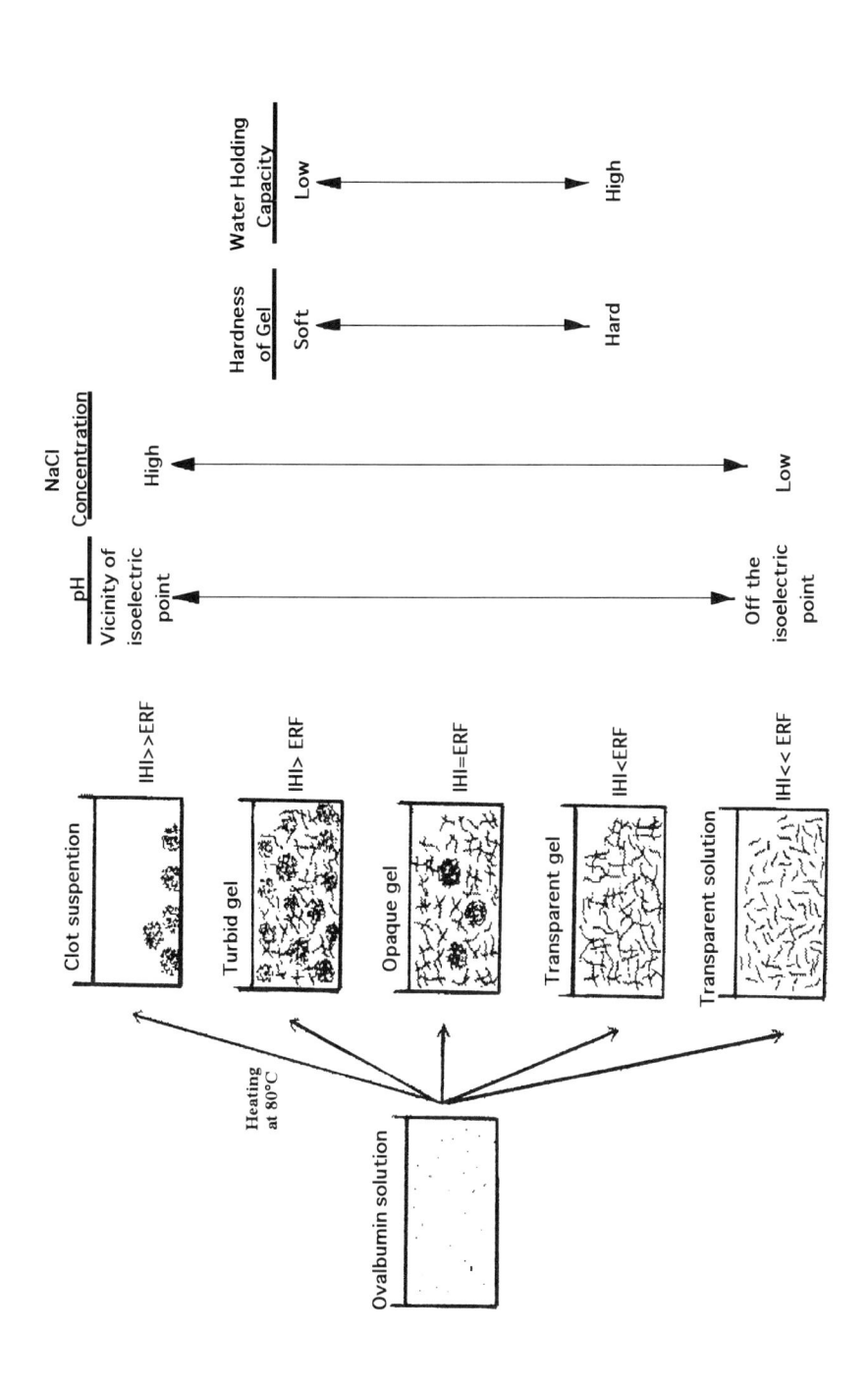

Figure 8. A schematic representation of several tangled states of ovalbumin irreversibly denatured by heating under various conditions. IHI, intermolecular hydrophobic interaction; ERF, electrostatic repulsive force.

3. Structural changes of ovalbumin molecules on heating

To make clear the change in the secondary structure of heat denatured OVA, Raman spectrum analysis [10] and circular dichroism (CD) analysis [11] were applied. The results revealed that a new β-sheet structure was produced in the denatured OVA. It has been reported that a correlation exists between the constituting ratio of β-sheet structure and the amount of soluble polymers of heat denatured OVA [11] (see Chapter 4). Some recent investigations have demonstrated that denaturation of many globular proteins proceeds via a state called the molten globular state [12, 13]. The molten globular state refers to a partially denatured protein which still preserves its globular shape. Koseki and his co-workers found that under low ionic strengths, the change in the secondary structure of OVA was small in CD and UV spectral analysis. Therefore they assumed that OVA remains in the molten globular state upon heating under low ionic strengths [9].

III. FOAMING PROPERTY

A. FORMATION OF FOAM AND ITS STABILITY

Foaming property is expressed as a surface activating ability of a solute which decreases the surface tension of the solvent. Substances having a surface activating ability gather on the surface of the solvent. These substances possess hydrophilic and hydrophobic groups in their molecules. The hydrophilic groups are arranged toward the water phase and the hydrophobic groups toward the air phase. When air comes into a solution to form bubbles, the surface activating substance adheres to the surface of bubbles producing films, and these bubbles gather to become foam. As time passes, the bubbles become weaker and finally break down, since the water begins to exude from the bubble (drainage). Foam stability is dependent on the force of supporting the film of the bubble. If the film of the bubbles are constructed with solid or solid-like substances, the foam is generally stable.

A solution of protein is a colloidal solution in which proteins exhibit the ability to decrease the surface tension of the solvent. Proteins generally produce stable foams, since proteins are denatured to become unfolded molecules which contribute to generating stable films. Damondaean suggested the following for a protein to display excellent foaming capacity [14]:

1) to be able to adhere quickly to the bubble surface,
2) to be able to change their structures rapidly and assemble together properly on the bubble surface, and
3) to be able to produce firm films by mutual interaction of the molecules.

B. FOAMING ABILITY OF EGG ALBUMEN
1. Method for evaluation of foaming ability

The foaming ability and foam stability of egg albumen are generally evaluated by the following method: The volume of foam and the amount of solution used to produce foam are employed as parameters of foaming ability. Also, the decrease in the volume of foam at certain time intervals and the amount of drainage are parameters of foam stability. In preparation of angel cake and meringue using egg albumen, the volume of the product and fineness of foams inside the product are also used to evaluate the foaming ability and foam stability of egg albumen.

2. Influence of freshness of egg albumen

During storage, the thick egg albumen changes into thin egg albumen decreasing its viscosity. In fresh hen eggs, 50% of egg albumen is thick, but it decreases to 30% after 12 days, storage at 25°C. We investigated the relationship between freshness of eggs (viscosity of egg albumen) and foam stability of the products obtained therefrom. The result revealed that a decrease in foam stability occurred with a decrease in the viscosity of egg white (Figure 9). It has been known that ovomucin and lysozyme are in a conjugated state to produce the structure of thick albumen, and the content of ovomucin in thick albumen is larger than that in thin albumen.

To investigate the foam stability of thick and thin albumen, the albumen from fresh eggs was separated into thick and thin albumen. They were mixed at several ratios and subjected to meringue tests. The results showed that the foam stability index positively correlated with the concentration of thick albumen (Figure 10), indicating that the viscosity produced by albumen directly affects the foam stability of meringue.

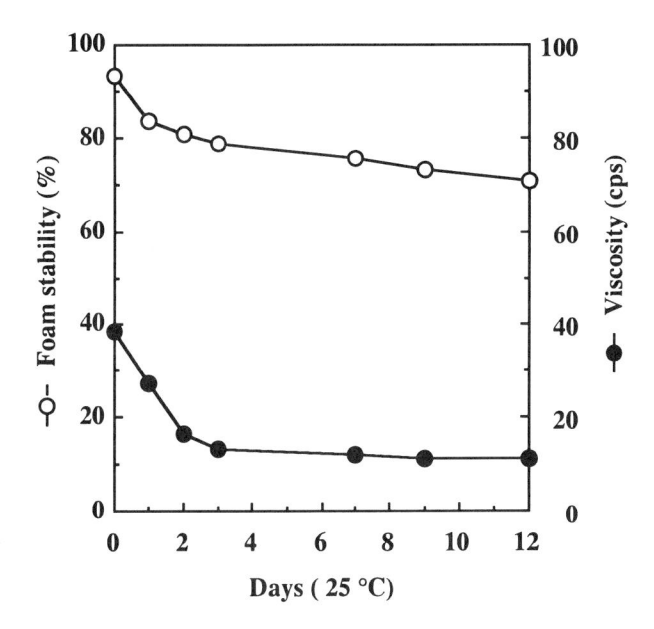

Figure 9. Influence of the storage period of eggs on viscosity and foam stability of egg albumen. Egg albumen (300 ml) was whipped until the specific gravity of the foam reached 0.15. Foam stability is expressed by the weight percentage of egg albumen in the foam after the foam was left for 30 min at 25°C.

3. Effect of pasteurization

Pasteurization of egg albumen decreases its foaming ability and results in reduction of the quality and volume of angel cakes. It has been reported that this phenomenon is due to denaturation of ovotransferrin on pasteurization. Many papers have reported on experiments for controlling the foaming ability of pasteurized egg albumen. Ovotransferrin denatures at 63°C. However, its denaturation temperature goes up to 84°C if ovotransferrin is associated with Fe, Cu, Al, or other metallic ions [15]. It is also known that addition of aluminum sulfate to egg albumen brings a rise in the temperature for the denaturation of ovotransferrin. This fact has been known to be one of the methods of controlling the foaming ability of egg albumen. Phosphoric and citric acids also have high affinity for ovotransferrin. Therefore the addition of their salts also increases the denaturation temperature of ovotransferrin, improving the foaming ability of egg albumen on pasteurization [16].

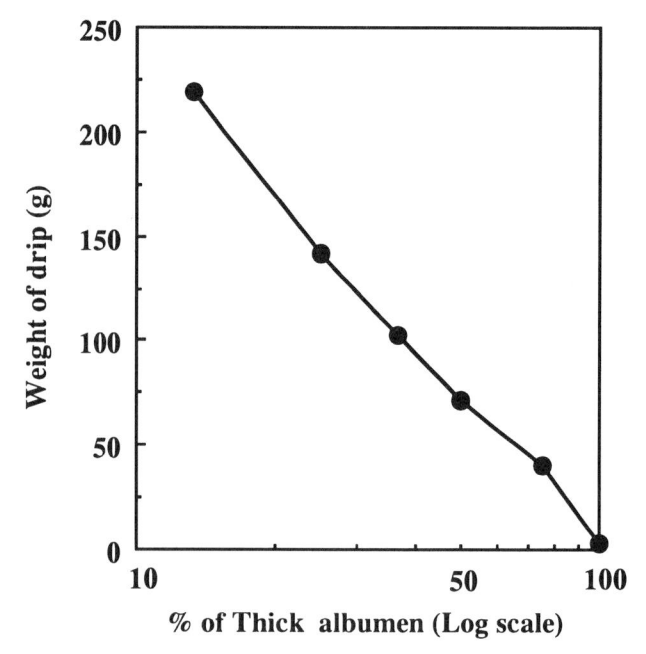

Figure 10. Influence of mincing proportion of thick egg albumen on foam stability in meringue test. Reconstituted egg albumen (300 g) was mixed with sugar (150 g), and then whipped until the specific gravity of meringue reached 0.2. Then, the meringue was allowed to stand for 3 hr at 25°C and the weight of drip was estimated as an indicator of foam stability.

4. Other factors affecting foaming of egg albumen

The addition of glycerol, sorbitol, or several other chemicals which increase the viscosity of egg albumen improve the foam stability of egg albumen [17]. In general, the use of egg albumen with a viscosity increased by the addition of any of the above chemicals, results in good stability of the foam produced but reduces the foaming ability. Therefore, adjusting the viscosity of egg albumen mixture to a moderate degree is necessary for obtaining good foaming ability of egg albumen.

Upon mixing egg yolk with egg albumen, the foaming ability decreases, because egg yolk lipids depress the formation of foam. It has been reported that, after removal of lipids by centrifugation or hydrolysis of lipids by enzyme treatment, the foaming activity of the mixture was recovered [18].

C. FOAMING ABILITY OF EGG WHITE PROTEINS

Johnson and Zabik examined the foaming properties of each protein of egg albumen in making angel cake. They found that the contribution of ovoglobulin to the volume of the angel cake is the biggest among various proteins in albumen. The structure and texture of the cake were fine and with very good quality. The foaming abilities of ovalbumin was not particularly high, but the cake volume was comparatively larger. The bubbles in the structure, however, were coarse and rough to be evaluated as bad quality products. The foaming abilities of ovotransferrin, lysozyme, ovomucoid, and ovomucin were evaluated as weak, and the cake volume was not good either [19]. MacDonell and his colleagues have also mentioned that the roles of globulin, ovomucin, and ovalbumin in angel cake production were foaming ability, foam stability, and the structure of the cake, respectively [20].

In a comparison of foaming capacity of egg white proteins, Nakamura and his co-workers reported that foaming ability was in the following decreasing order, with considerable differences: globulin > ovotransferrin > ovomucoid > ovalbumin > lysozyme [21]. It was presumed that the more sensitive to heat denaturation, the more effective is the protein in foaming ability. The values of pH greatly influence the foaming ability. On the extremely acidic or alkaline side or at the region of the isoelectric point, ovalbumin and ovotransferrin show high foaming ability [21]. The proteins showing high foaming ability at high pH values are generally sensitive to denaturation. This tendency is observed to be accompanied by intermolecular interactions of the denatured proteins.

D. MECHANISM OF FOAMING OF OVALBUMIN

In the study of foaming ability of proteins in regard to their hydrophobicity, Li-Chan and Nakai reported that there is a positive relationship between the degree of hydrophobicity increased by heat denaturation and the foaming ability of the proteins [22]. Kato reported that a remarkable increase of foaming ability and foam stability of ovalbumin were observed with an increase in the hydrophobicity of ovalbumin molecules by heat denaturation [23].

In general, ovalbumin exhibits low foaming ability. However, we observed that, after heat denaturation, the foaming ability becomes 12 times higher than that of the native ovalbumin. This might be because the heat denatured ovalbumin molecules are liable to interact mutually, resulting in strong foaming ability. Doi and his co-workers reported that ovalbumin in foams forms -S-S- bonds. However, the formation of foams occurs regardless of the presence or absence of reducing agents, indicating that the formation of -S-S- bonds is not necessarily related to foam formation [24].

IV. EMULSIFICATION

A. EMULSIFICATION AND EMULSIFIED FOODS

Oil and water never mix uniformly, unless certain conditions are provided. If one of them is dispersed uniformly into the other, the solution obtained is called an emulsion. Emulsifiers have hydrophobic (oil affinity sites) and hydrophilic groups (water binding sites), and are used to produce emulsion. These groups adhere to oil, on one hand, and water molecules, on the other hand. These properties of an emulsifier are important for obtaining a stable emulsion. There are two types of emulsions, oil in water (O/W) and water in oil (W/O). An O/W emulsion has electric conductivity whereas a W/O emulsion does not show this property.

Many emulsified foods, such as ice cream, whipped cream, coffee cream, margarine, mayonnaise, and dressings are now available almost everywhere. It is noticeable that many popular emulsified foods are made by using egg yolk. The emulsifying activity of whole egg and egg albumen are 1/2 and 1/4 of that of egg yolk alone, respectively [25]. Egg yolk is known as one of the best emulsifiers used in food processing.

B. CHARACTERISTICS OF EGG YOLK AS AN EMULSIFIER
1. Method of evaluation of emulsifying properties

The emulsifying properties of food proteins are evaluated for their emulsifying capacity and the stability of the emulsion produced. To evaluate the emulsifying capacity, oil is added little by little to a definite amount of an emulsifier-containing solution, and the amount of oil required for transition from an O/W to a W/O emulsion is determined. To evaluate the stability of the emulsion prepared, the amount of oil or water separated from the emulsion is determined after leaving the emulsion under certain established conditions.

2. Effects of freezing of egg yolk

Freezing of egg yolk causes a gradual denaturation of yolk proteins, accompanied by a gradual increase in viscosity. In addition, the process of freezing and thawing results in irreversible gelation of egg yolk. It is known that freezing of egg yolk decreases its emulsifying capacity and the stability of the emulsion [26]. Our experimental results carried out with the egg yolk pasteurized and then frozen at -25°C for a week, showed that the emulsifying capacity decreased to 67% of that of the original egg yolk (Table 1). In the following 20 weeks of storage, however, the emulsifying capacity was maintained without further loss (62%). We also observed that the addition of 10% sugar or salt to the egg yolk to be stored frozen protects it from deterioration. In manufacturing mayonnaise and dressings, egg yolk with 10% of NaCl added has been widely used.

Table 1
Changes in Emulsifying Activity of Egg Yolk Stored in a Frozen State

	Control	1	2	5	10	20
Corn oil (g)	920	615	630	583	610	570

Storage periods of egg yolk at -25°C (weeks)

Sterilized egg yolk (62.5°C, 3.5 min.) stored at -25°C was used. The control was unfrozen but sterilized egg yolk.. A 30 g egg yolk was used for the test after adding 2% NaCl and 10% vinegar by weight, and the corn oil (g) required for causing phase inversion of the o/w emulsion was determined.

3. Effect of pasteurization

It has been reported that egg yolk pasteurized at 62°C for 5 min exhibits higher emulsifying capacity than unpasteurized yolk. However, pasteurization at higher temperatures causes a slight decrease in emulsifying capacity [27]. Tsutsui reported that the emulsifying capacity and emulsion stability of low density lipoproteins (LDL, used as a 4% solution), were not affected by sterilization at 75°C for 5 min, but at 80°C, these abilities were lost rapidly [28].

4. Effect of acids or salts

The emulsion stability of egg yolk is increased by the addition of salt, but the addition of acetic acid decreases the emulsifying capacity [29]. Table 2 shows the relation between the concentrations of acetic acid added and the emulsifying capacity of egg yolk in a test of making mayonnaise. High concentrations of acetic acid resulted in reduction of emulsifying capacity even in the presence of a high concentration of NaCl.

5. Effects of drying egg yolk

Drying of egg yolk decreases emulsifying capacity [30]. This is because the lipoproteins release lipids during the drying process. Consequently, the emulsifying capacity of the yolk powder is inferior to that of the original egg yolk. The addition of sucrose (5-10%) to egg yolk to be dried is effective for the production of egg yolk powder whose emulsifying activity is comparable to that of the original egg yolk [31].

Table 2
Emulsifying Activity of Egg Yolk as a Function of NaCl and Vinegar Concentration

		Vinegar %					
		0	1	2	3	4	5
NaCl %	2	1000	1030	1100	1100	1060	1000
	4	1050	950	940	770	800	900
	6	1100	1020	960	750	750	780
	8	1120	1050	950	700	720	750
	10	1100	1030	950	750	730	725

The weight of corn oil required for causing phase-reversion of o/w emulsion was expressed as the emulsifying activity of egg yolk. Corn oil (g) was added to 30 g of sterilized (62.5°C, 3.5 min) egg yolk containing various concentrations of NaCl and vinegar until phase inversion of the o/w emulsion was achieved.

C. EMULSIFYING PROPERTIES OF EGG YOLK LIPOPROTEINS

Egg yolk is an excellent emulsifier used in making mayonnaise and other products. The high emulsifying capacity of yolk depends on the properties of the LDL [26]. Low density lipoprotein (LDL) is a major component of yolk plasma whereas high density lipoprotein (HDL) is a component of yolk granules. Oshida reported that the emulsifying capacity of LDL was greater than that of HDL. However, in the emulsion stability test, the result was reversed [32].

The emulsifying capacity of proteins is closely related to the structure of the protein, especially its hydrophobicity [33, 34]. Bovine serum albumin (BSA) rich in hydrophobicity is said to have excellent emulsifying ability. Mizutani and Nakamura reported that the emulsifying capacity of LDL was considerably greater than that of BSA. They mentioned that LDL adheres to oil drops faster than BSA, forming fine oil particles. As for emulsion stability, LDL was superior to BSA. They commented that the excellent emulsifying properties of LDL were not only due to the effect of the structure of the protein and phospholipids in LDL, but also to the LDL structure complexed with proteins (phosvitin) and lipids [35-37].

REFERENCES

1. **Donovan, J. W., Mapes, C. J., Davis, J. G., and Garibaldi, J. A.**, A differential scanning calorimetric study of the stability of egg white to heat denaturation, *J. Sci. Food Agric.*, **26**, 73, 1975.
2. **Holt, D. L., Watson, M. A., Dill, C. W., Alford, E. S., Edwards, R. L., Diehl, K. C., and Gardner, F. A.**, Correlation of the rheological behavior of egg albumen to temperature, pH, and NaCl concentration, *J. Food Sci.*, **49**, 137, 1984.

3. **Ogawa, N. and Tanabe, N.**, The scanning electron micrographs of the hard cooked albumen obtained from fresh and aged shell eggs of the chicken, *Jap. Poult. Sci. (in Japanese)*, **27**, 426, 1990.
4. **Sato, Y. and Nakamura, R.**, Functional properties of acetylated and succinylated egg white, *Agric. Biol. Chem.*, **41**, 2163, 1977.
5. **Seideman, W. E., Cotterill, O. J., and Funk, E. M.**, Factors affecting heat coagulation of egg white, *Poult. Sci.*, **42**, 406, 1963.
6. **Wang, A. C., Funk, K., and Zabik, M. E.**, Effect of sucrose on the quality characteristics of baked custards, *Poult. Sci.*, **53**, 807, 1974.
7. **Hatta, H., Kitabatake, N., and Doi, E.**, Turbidity and hardness of a heat-induced gel of hen egg ovalbumin, *Agric. Biol. Chem.*, **50**, 2083, 1986.
8. **Kitabatake, N., Hatta, H., and Doi, E.**, Heat-induced and transparent gel prepared from hen egg ovalbumin in the presence of salt by a two-step heating method, *Agric. Biol. Chem.*, **51**, 771, 1987.
9. **Koseki, T., Fukuda, T., Kitabatake, N., and Doi, E.**, Characterization of linear polymers induced by thermal denaturation of ovalbumin, *Food Hydrocoll.*, **3**, 135, 1989.
10. **Painter, P. C. and Koenig, J. L.**, Raman spectroscopic study of the proteins of egg white, *Biopolymers*, **15**, 2155, 1976.
11. **Kato, A. and Takagi, T.**, Formation of intermolecular β-sheet structure during heat denaturation of ovalbumin, *J. Agric. Food Chem.*, **36**, 1156, 1988.
12. **Pitsyn, O. B.**, Protein folding: Hypotheses and experiments, *J. Protein Chem.*, **6**, 273, 1987.
13. **Kuwajima, K., Mitani, M., and Sugai, S.**, Characterization of the critical state in protein folding. Effects of guanidine hydrochloride and specific Ca^{2+} binding on the folding kinetics of α-lactalbumin, *J. Mol. Biol.*, **206**, 547, 1989.
14. **Damodaean, S.**, *Protein Functionality in Food Systems*, Hettarachchy, N. S. and Ziegler, G. R., Eds., Marcel Dekker, New York, 1994.
15. **Cunningham, F. E. and Lineweaver, H.**, Stabilization of egg-white proteins to pasteurizing temperatures above 60 °C, *Food Technol.*, **19**, 136, 1965.
16. **Nakamura, R., Umemura, O., and Takemoto, H.**, Effect of heating on the functional properties of ovotransferrin, *Agric. Biol. Chem.*, **43**, 325, 1979.
17. **Nakamura, R. and Sato, Y.**, Studies on the foaming property of the chicken egg white. On the coagulated proteins under various whipping conditions, *Agric. Biol. Chem.*, **28**, 524, 1964.
18. **Cotterill, O. J. and Funk, E. M.**, Effect of pH and lipase treatment on yolk-contaminated egg white, *Food Technol.*, **17**, 103, 1963.
19. **Jhonson, T. M. and Zabic, M. E.**, Egg albumen proteins interactions in an angel food cake system, *J. Food Sci.*, **46**, 1231, 1981.
20. **Macdonnell, L. R., Feeney, R. E., Hanson, H. L., Campbell, A., and Sugihara, T. F.**, The functional properties of the egg white proteins, *Food Technol.*, **9**, 49, 1955.
21. **Nakamura, R. and Sato, T.**, Studies on the foaming property of the chicken egg white: Whipping properties of various egg white proteins, *Nippon Nogeikagaku Kaishi (in Japanese)*, **35**, 385, 1961.
22. **Li-Chan, E. and Nakai, S.**, Effects of molecular changes (SH groups and hydrophobicity) of food proteins on their functionality, in *Food Proteins*, Kinsella, J. E. and Soucie, W. G., Eds., American Oil Chemist's Society, Champaign, IL, USA, 1989, p 232.
23. **Kato, A., Komatsu, K., Fujimoto, K., and Kobayashi, K.**, Relationship between surface functional properties and flexibility of proteins detected by the protease susceptibility, *J. Agric. Food Chem.*, **33**, 931, 1985.

24. **Doi, E., Kitabatake, N., Hatta, H., and Koseki, T.,** Relationship of SH groups to functionality of ovalbumin, in *Food Proteins*, Kinsella, J. E. and Soucie, W. G., Eds., American Oil Chemist's Society, Champaign, IL, USA, 1989, p 252.
25. **Lowe, B.,** Egg, in *Experimental Cookery*, John Wiley & Sons, Inc., New York, 1964.
26. **Davey, E. M., Zabik, M. E., and Dawson, L. E.,** Fresh and frozen egg yolk protein fractions: Emulsion stabilizing power, viscosity, and electrophoretic patterns, *Poult. Sci.*, **48**, 251, 1969.
27. **Cotterill, O. J., Glauert, J., and Bassett, H. J.,** Emulsifying properties of salted yolk after pasteurization and storage, *Poult. Sci.*, **55**, 544, 1976.
28. **Tsutsui, T.,** Functional properties of heat-treated egg yolk low density lipoprotein, *J. Food Sci.*, **53**, 1103, 1988.
29. **Oshida, K.,** Basic studies on mayonnaise manufacturing: Part III. Effects of sodium chloride and acetic acid on stability of mayonnaise, *Nippon Shokuhin Kogyo Gakkaishi (in Japanese)*, **22**, 501, 1975.
30. **Zabik, M. E.,** Comparison of frozen, foam-spray-dried, freeze-dried, and spray-dried eggs. VI. Emulsifying properties at three pH levels, *Food Technol.*, **23**, 838, 1969.
31. **Schitz, J. R., Snyder, H. E., and Forsythe, R. H.,** Co-dried carbohydrates effect on the performance of egg yolk solids, *J. Food Sci.*, **33**, 507, 1968.
32. **Oshida, K.,** Basic studies on mayonnaise manufacturing: Part IV. Effect of sodium chloride and acetic acid on emulsifying capacity and emulsifying stability of egg yolk's low-density and high-density fractions, *Nippon Shokuhin Kogyo Gakkaishi (in Japanese)*, **23**, 250, 1976.
33. **Nakai, S.,** Structure-function relationships of food proteins with an emphasis on the importance of protein hydrophobicity, *J. Agric. Food Chem.*, **31**, 676, 1983.
34. **Kato, A. and Nakai, S.,** Hydrophobicity determined by a fluorescence probe method and its correlation with surface properties of proteins, *Biochim. Biophys. Acta*, **624**, 13, 1980.
35. **Mizutani, R. and Nakamura, R.,** Emulsifying properties of egg yolk low density lipoprotein (LDL): Comparison with bovine serum albumin and egg lecithin, *Lebensm. -Wiss. u. -Technol.*, **17**, 213, 1984.
36. **Mizutani, R. and Nakamura, R.,** Physical state of the dispersed phases of emulsions prepared with egg yolk low density lipoprotein and bovine serum albumin, *J. Food Sci.*, **50**, 1621, 1985.
37. **Mizutani, R. and Nakamura, R.,** The contribution of polypeptide moiety on the emulsifying properties of egg yolk low density lipoprotein (LDL), *Lebensm. -Wiss. U. -Technol.*, **18**, 60, 1985.

Chapter 9

ENZYMES IN UNFERTILIZED HEN EGGS

A. Seko, L.R. Juneja, and T. Yamamoto

TABLE OF CONTENTS

I. Introduction
II. Enzymic Activities in Hen Eggs
    A. Proteolytic Enzymes
    B. Phosphatases
    C. Glycosidases
    D. Nucleoside Triphosphatases
    E. Ribonucleases (RNases)
    F. Ribonucleic Acid-degrading Enzymes
    G. Pyruvate Kinases and Glycolytic Enzymes
    H. Other Enzymes
III. Discussion

References

I. INTRODUCTION

The study of enzymic activities found in hen eggs had started in the early years of this century and have become refined and steady in relation to the development of enzyme kinetics and protein chemistry. About 30 kinds of enzymic activities have so far been found in unfertilized hen eggs and, in some cases, enzymes have been purified to homogeneity and the biochemical, enzymological properties, and, even if to a lesser extent, the biological significance clarified.

In early works, rather weak activities of enzymes in a large amount of egg yolk or egg white had often made the studies difficult for identification and for excluding bacterial contamination in the reaction systems. Lineweaver and his colleagues (1948) reinvestigated the egg enzymes which had been reported before, especially the quantification of enzymic activities and antibacterial reagents added during incubation, and they found tributyrinase (esterase), peptidase, and catalase in yolk and egg white, and amylase and phosphatase only in yolk within the limits examined [1]. Thereafter, the existence, localization, and enzymological properties of various types of enzymes have been clarified and, in some cases, enzymes have been purified. Enzymes studied in unfertilized eggs so far are listed in Table 1. The comments about these studies are outlined below. Since lysozyme, the most well known enzyme in hen eggs, is described in Chapter 4, it is not mentioned in this chapter.

II. ENZYMIC ACTIVITIES IN HEN EGGS

A. PROTEOLYTIC ENZYMES

After the study by Lineweaver and his colleagues (1948), acid proteases and amino peptidases have been investigated well. First, two acid proteases, S-prot 1 and S-prot 2, were purified, from yolk supernatant (S), and from yolk granules (G), two acid proteases, G-prot 1

Table 1
Enzymic Activities Found in Unfertilized Hen Eggs

Enzyme	Source	Reference
β-N-acetylglucosaminidase	vitelline membrane, white	20-22, 41-43
Acid phosphatase	vitelline membrane, white, yolk	12-15, 44, 45
Alkaline phosphatase	vitelline membrane, white, yolk	1, 12, 13, 16-19, 28, 44-49
Aminopeptidase	white, yolk	5-11
Amylase	yolk	1, 40, 50
ATPase	plasma + vitelline membrane	24-26
Catalase	white, yolk	1, 21
Cholinesterase	yolk	28, 38, 39, 51, 52
Deaminase	yolk	28, 33, 34
Deoxyribonuclease	yolk	28
β-D-galactosidase	yolk	28
α-D-glucosidase	yolk	28
β-D-glucosidase	yolk	28
β-D-glucuronidase	yolk	28
Glutamate oxaloacetate transaminase	yolk	28
Glyceraldehyde-3-phosphate dehydrogenase	vitelline membrane, yolk	36
Hexokinase	vitelline membrane, yolk	36
Lactate dehydrogenase	vitelline membrane, yolk	36
Lysozyme	white	53
α-Mannosidase	white, yolk	20, 21, 23, 28
Nucleosidase	yolk	33
Nucleoside triphosphatase	vitelline membrane, white, yolk	27, 46
5'-Nucleotidase	vitelline membrane, yolk	54, 55
Peptidase	white, yolk	28, 46, 49
Protease	yolk	1, 3, 4, 56, 57
Purine N_1-C_6 hydrolase	yolk	35
Pyruvate kinase	vitelline membrane, yolk	36, 37
Ribonuclease	white, yolk	28-32, 58
Tributyrinase	white, yolk	1

Partially cited from Burley and Vadehra (1989) [2] by permission of John Wiley & Sons, New York.

and G-prot 2, were also isolated by DEAE-cellulose chromatography, gel filtration, and pepstatin-Sepharose affinity chromatography [3, 4]. Prot 1 series were eluted at lower ionic strength from DEAE-cellulose than prot 2 series. Optimal pHs of these proteases are all around 3.0. The difference between prot 1 series and prot 2 series is evident from several points. Prot 1 series are unstable at pH 2, more resistant to p-bromophenacylbromide (p-BPB) and diacetyl-norleucine-methylester/Cu^{2+} complex (DAN/Cu^{2+}), hydrolyze insulin better, and do ovalbumin, myoglobin, and phosvitin less than prot 2 series. Apparent molecular weight was estimated as 50 K (S-prot 1), 40 K (S-prot 2), 47 K (G-prot 1) and 38 K (G-prot 2). The pH profile and the action on several proteins or artificial substrates suggested that these four enzymes are cathepsin D-like proteinases; Wouters and his colleagues (1985) named S-prot 1 and G-prot 1 as yolk-cathepsin D1, and S-prot 2 and G-prot 2 as yolk-cathepsin D2 [4].

Aminopeptidases have also been found in both yolk [5] and egg white [6]. Ichishima *et al.* purified aminopeptidase, Ey, from egg yolk by ammonium sulfate fractionation followed by applying chromatography on several columns to 13,000-fold [5]. The molecular weight of the enzyme is 330 K and it consists of 150 K of homodimer [7]. The isoelectric point is 2.8; this low value was explained partly by the existence of sialic acid (4.8 % w/w) and thus, this enzyme, which additionally contains GlcNAc (2.8 %), GalNAc (0.36 %), and hexose (6.0%), is a glycoprotein [8]. The optimal pH is 6.5 and Michaelis constant (Km) is 0.01 mM for leucine-4-methylcoumaryl-7-amide as a substrate. The catalytic activity needs metal ions because of the loss of activity in the presence of bestatin, o-phenanthroline, and EDTA [5]. Evidences also support the aminopeptidase Ey as a zinc-metalloenzyme because the enzyme contains 1.1 g atom of per/mol of the enzyme subunit as quantified by atomic absorption and because the apoenzyme prepared by EDTA treatment is reactivated by the addition of divalent cations, Zn^{2+}, Co^{2+}, Cu^{2+}, Ca^{2+}, and so on [8]. The N-terminal amino acid sequence was determined as acyl-Xaa-Xaa-Pro-Glu-Ala-Ala-Ser-Leu-Pro-Gly- by Edman degradation carried out after reaction with an acyl amino acid releasing enzyme [8]. With respect to substrate specificity, this enzyme was considered to be an aminopeptidase and can hydrolyze almost all kinds of N-terminal amino acids in substrate peptides, regardless of the presence of electric charges or hydrophobicity, and, in some cases, even N-terminal Xaa-Pro-sequences on which other aminopeptidases are generally unable to act [9]. N-terminal Xaa-Pro-sequences in tetra- or pentapeptides can be hydrolyzed, but those in longer or shorter peptides cannot. In the case of Pro-Phe-Gly-Lys, the N-terminal proline residue can be hydrolyzed, but the proline residue in shorter or longer peptides can not. α-NH_2-acetylated amino acid residues, pyroglutamic acid, and sarcosine (N-methyl-glycine) at N-termini can not be hydrolyzed, buta an N-formylmethionyl residue in longer than dipeptides can be released. The release of formyl group was not observed. From the ability of aminopeptidase Ey to act on N-formylmethionine peptides, Tanaka and Ichishima speculated that the enzyme could play a role as an anti-bacterial factor in inactivating bacterial N-formylpeptides having some unfavourable effects such as chemotaxis during hen's embryogenesis [10].

Two types of aminopeptidases, glutamylaminopeptidase [6] and methionine-preferring aminopeptidase, have been found in egg white. Petrovic and Vitale (1990), and Skrtic and Vitale (1994) [11] purified these peptidases from egg white by ammonium sulfate fractionation followed by chromatography on several ion exchangers to 100,000 and 45,000-fold, respectively. The former was finally obtained 90.2 mU per egg using α-Glu-4-nitroanilide as a substrate and the latter, 323 U per egg using leucine-2-naphthylamide (2NA). The enzymes have following properties; first, the glutamylaminopeptidase has MW 320 K, which consists of 180 K of homodimer, and has glycan chain(s) because of absorption on a Con A-Sepharose column [6]. The isoelectric point is 4.2 and the optimal pH is 7.6 for α-Glu-2NA as a substrate. The activity is strongly inhibited by EDTA or N-bromosuccinimide, and activated by Mn^{2+} or Ca^{2+}. The enzyme can hydrolyze only acidic amino acid residues such as α-Glu-2NA, α-Asp-2NA, Asp-Ala, and so on. Second, methionine-preferring aminopeptidase has MW 180 K, and is strongly inhibited by o-phenanthroline [11]. The optimal pH is 7.0-7.5, Met-2NA, Ala-2NA, and Leu-2NA are good substrates, but acidic amino acid-2NAs and N-terminal formyl amino acids can not be hydrolyzed. The enzyme shows strong affinity for long peptides.

B. PHOSPHATASES

In this section, phosphatases other than ATPase will be mentioned. Moors and Stockx have observed the activities of acid and alkaline phosphatases in egg yolk [12, 13]. Debruyne partially purified the acid phosphatase from egg yolk by delipidation, gel filtration, and hydrophobic column chromatography [14]. The enzymic activity recovered from 115 ml of egg yolk was 98.8 mU examined on p-nitrophenylphosphate as substrate. The purification factor was 341-fold. Apparent MW is 31.4 K and the optimal pH, 5.1. This enzyme easily associates depending on ion strength. 5'-AMP, 3'-AMP, phosphoryl choline, and creatine phosphate are good substrates whereas o-phosphoserine, o-phosphothreonine, phosphoethanolamine, glucose-6-phosphate, 5'-ADP, and phosvitin are not hydrolyzed. Inhibitory effects were observed for pyrophosphoric acid, Na_2SO_4, NaF, dihydroxymalonic acid, and hydroxymalonic acid, but only weakly for α-hydroxycarboxylic acids known as inhibitors for acid phosphatase. The enzyme was shown to have transphosphorylation activity capable of transferring a phosphate group to a hydroxyl group of n-butanol, ethyleneglycol, glycerol, and dihydroxyethylether at high concentrations of them [14]. Acid phosphatase found in vitelline membrane can be solubilized with 3% Tween 80 [15]. 2'-, 3'-, 5'-Monophosphonucleotides and 5'-monophosphodeoxynucleotides are good substrates whereas o-phosphoserine, o-phosphothreonine, phosphoryl choline, and phosphoryl ethanolamine are poor. EDTA has no inhibitory effect on this enzyme. Acid phosphatase activity was also found in egg white showing an optimal pH at 4.9 [12, 13].

The alkaline phosphatases have been found in vitelline membrane, yolk, and egg white [12, 13]. De Boeck and Stockx (1978) have found alkaline phosphatase activity from a vitelline membrane fraction whose optimal pH was 9.6 [16]. This enzyme can hydrolyze the phosphate ester linkages of 2'-, 3'-, 5'-nucleotides, o-phosphoserine, o-phosphothreonine, and phosvitin-derived peptides containing two to five successive phosphoserine residues, although the enzyme cannot liberate phosphate from intact phosvitin. The well-known inhibitors of alkaline phosphatases, L-phenylalanine, L-homoarginine, tetramisole, and (-)p-bromotetramisole, inhibit the enzyme, and EDTA also inhibits it. Ionic detergents, deoxycholate and cholate, enable the solubilization of the enzyme associated with vitelline membrane. Debruyne and Stockx (1979) purified an alkaline phosphatase from egg yolk by delipidation followed by applying sequential chromatographies [17]. Alkaline phosphatase amounting to 63.3 U was recovered from 150 ml of yolk/water=1:1 fraction examined on p-nitrophenylphosphate as a substrate. The purification factor was 43,600-fold. The molecular weight is 150 K consisting of two 68.8 K subunits. The enzyme was considered to have heterogeneity and glycan chain(s) by the fact that on isoelectric focusing it was divided into three components and on a Con A-Sepharose column it was separated into two fractions. It was suggested that the yolk alkaline phosphatase, likewise vitellogenin, might be synthesized in the liver and accumulated in the yolk through the blood stream, since the isolated enzyme was similar to liver and blood alkaline phosphatases with respect to the elution pattern on anion-exchange chromatography and the molecular weight. Debruyne (1982) further studied the enzymological properties of yolk alkaline phosphatase [18]. This enzyme acts on o-phosphoserine, o-phosphothreonine, and ATP, but not on phosvitin. The optimal pH is about 8-10, which is strongly affected by the substrate used and substrate concentration. EDTA treatment causes the loss of activity, and conversely, the addition of Mg^{2+} activates the enzyme. The alkaline phosphatase, likewise acid phosphatase described above, has transphosphorylation activity at an alkaline pH [19]. Aliphatic alcohols having α-amino or α-imino groups are good acceptors of phosphate groups to their hydroxyl groups. The transphosphorylation was independent of leaving groups of donor molecules, thereby, the existence of an enzyme-phosphate intermediate(s) was predicted. Moreover, various kinetic studies suggested that there are at least two enzyme-phosphate intermediates and that one, produced first immediately after releasing a donor molecule lacking a phosphate group, can transfer a phosphate group to an acceptor alcohol, and then isomerizes to the other which can transfer phosphate to water. In egg white, too, alkaline phosphatase activity with optimal pH of 9.5 was found [12, 13].

C. GLYCOSIDASES

The most well-known and relatively large component of glycosidase in hen eggs is lysozyme, which was mentioned in Chapter 4. Other glycosidases so far found are β-*N*-acetylglucosaminidase, β-D-galactosidase, α-D-glucosidase, β-D-glucosidase, β-D-glucuronidase, and α-mannosidase. In 1966, Lush and Conchie found β-*N*-acetylglucosaminidase and α-mannosidase activities in hen egg albumen and showed that β-*N*-acetylglucosaminidase activity decreases rapidly during preservation of unfertilized eggs at room temperature [20]. It was shown that α-mannosidase was not inactivated by heat treatment at 60°C for 2.5 min whereas β-*N*-acetylglucosaminidase loses activity rapidly under the same conditions in spite of being stable at 53°C for 2.5 min [21]. This result indicates that β-*N*-acetylglucosaminidase activity can be used as a parameter for heat pasteurization of egg white [21, 22]. Thereafter, α-mannosidase was partially purified from a yolk soluble fraction by DEAE-cellulose and then by Sephadex G-200 gel filtration after delipidation with ether [23]. The optimal pH is 4.7 and apparent molecular weight is 250 K. Elucidation of biochemical and enzymological properties of these enzymes will be future subjects.

D. NUCLEOSIDE TRIPHOSPHATASES

Nucleoside triphosphatases including ATPase have been found in vitelline membrane and egg white. The enzyme derived from vitelline membrane requires Mg^{2+} and decreases its activity by the addition of EDTA or replacement of Mg^{2+} to Ca^{2+} [24-26]. ATPase of a membrane-associated form can be solubilized by repeated washing of a membrane fraction with relatively high concentrations of buffered KCl [25]. In membrane-associated form, K^+ is an activator, whereas in soluble form, it has no effect or rather a slightly inhibitory one. The effect of Na^+ is contradictory; it is an activator [26], or it has no effect [27]. The optimal pH was observed showing two peaks at 5.0-6.0 and 8.6 by Etheredge and his colleagues (1971) and one peak at 5.2-6.3 by Debruyne and Stockx (1978): These differences may be explained by the result that the latter authors observed that alteration of the pH profile depends on preparation methods for the enzyme [26, 27]. The apparent molecular weight is 260 K, which was determined using enzyme from vitelline membrane solubilized by sodium deoxycholate [27]. Debruyne and Stockx (1978) have purified a nucleoside triphosphatase from thick white albumen by gel filtration and lyophilization followed by extraction with sodium deoxycholate [27]. The molecular weight determined was 260 K showing an optimal pH at 8.0 before detergent extraction and 6.15 after this treatment. The activity increased in the presence of Mg^{2+} or Ca^{2+} and was diminished by EDTA. Although nucleoside triphosphatases from both sources show broad substrate specificity, nucleoside monophosphates are not good substrates and pyrophosphoric acid is not hydrolyzed, but it is a strong competitive inhibitor. Debruyne and Stockx (1978) suggested that these enzymes seem to have no relation both to transport ATPases because of weak inhibition by ouabain and no activating effects of Na^+ and K^+ on the egg enzymes and to contraction-triggering ATPases because Mg^{2+} is a better activator for the egg enzymes than Ca^{2+} [27].

E. RIBONUCLEASES (RNases)

Enzyme activities were observed in yolk [28] and egg white [29]. Wouters and Stockx (1977) found two RNase activities in egg yolk and partially purified them by chromatography on hydroxyapatite and a glass bead column following the solubilization from yolk granules; most of the yolk RNase activities are associated with yolk granules [30]. Their apparent molecular weight are 9.5 K and 11.3 K [31]. They have optimal pHs around neutral, are inactive toward DNA, and are inhibited by EDTA. Limited to dinucleotides as substrates, yolk RNases preferably act on pyrimidine nucleotides, whereas egg white RNase shows broad substrate specificity [32].

F. RIBONUCLEIC ACID-DEGRADING ENZYMES

In 1972, De Boeck and Stockx found a nucleosidase in yolk which released purine or pyrimidine bases from nucleosides, namely, *N*-glycosidase [33]. The optimal pH of the nucleosidase is 7.0 [33]. They also found an activity which deaminates adenosine, deoxyadenosine, guanosine, and cytidine. The deaminase was partially purified by delipidation and gel filtration. The optimal pH of the deaminase is 6.5, and the apparent molecular weight was about 14 K [34]. Unlike many other yolk enzymes, this enzyme did not show any aggregative nature. Moreover, they found an activity to cleave the N_1-C_6 bond of purine in the yolk fraction after the Sephadex G-75 gel filtration [35]. By this activity, xanthine was modified to 4-ureido-imidazole-5-carboxylic acid and 4-ureido-imidazole. The latter compound was thought to be produced from the former by spontaneous decarboxylation. The optimal pH is 7.0. The fact that, in yolk and egg white, no activity of xanthine oxidase, uricase, or allantoinase have been detected whereas unusual purine N_1-C_6 hydrolase was observed, may indicate that the reactions not using O_2 or coenzymes are favored in unfertilized eggs [35].

G. PYRUVATE KINASES AND GLYCOLYTIC ENZYMES

Pyruvate kinase catalyzes the reaction from phospho*enol*pyruvate and ADP to pyruvate and ATP, and it plays one of the most important roles in the regulation of glycolysis. The isozymes, type M, type L, and type A (or K), have been distinguished from each other and were shown to be localized in different tissues. Isozyme type M2, found in embryonal cells and tumor cells, is thought to be a key enzyme for sugar metabolism in proliferating cells. In 1986, Noda and Schoner studied the distribution of pyruvate kinase types M1 and M2, hexokinase, glyceraldehyde-3-phosphate dehydrogenase, and lactate dehydrogenase in unfertilized eggs [36]. The major portion of glycolytic enzymes is localized in the yolk fraction and the rest in the vitelline membrane, although the specific activities of those enzymes are higher in the vitelline membrane than in the yolk. As for pyruvate kinases, both types M1 and M2 are present in the yolk. However, in blastodisc, latebra prepared by cutting from lyophilized egg balls, and vitelline membrane, only type M2 exists. Egg yolk type M2 was indeed activated by the well-known specific effectors, fructose 1,6-bisphosphate and L-serine and inhibited by L-alanine, whereas egg yolk type M1 was never affected by these compounds. The presence of type M2 in the blastodisc is concomitant with the result that type M2 has so far been found in various proliferating or undifferentiated cells.

Noda and his colleagues (1986) have shown that egg yolk type M2 pyruvate kinase is phosphorylated by chicken oviduct-derived protein kinase [37]. Type M1 and M2 were purified from yolk both to about 10,000-fold. The activities recovered from one egg were 47.5 U and 198 U for type M1 and M2, respectively. Type M2 was also purified 1,100-fold from vitelline membrane and a preparation of 171 U activity was obtained from 300 eggs. Enzymic activities were determined using phospho*enol*pyruvate and ADP as substrates. Vitelline membrane-derived type M2 has a 250 K molecular weight. The enzyme may be a tetramer consisting of 60-62 K subunits. Purified type M2 was phosphorylated by protein kinase C derived from chicken oviduct while type M1 from egg yolk was also done, but weakly. The phosphorylation site of type M2 was found largely at serine and weakly at threonine, in analysis by electrophoresis following acid hydrolysis. The affinity of type M2 to phospho *enol*pyruvate was increased by phosphorylation. Noda and his colleagues (1986) speculated that the pyruvate kinase isozyme of type M2, potentially regulated by protein kinase C, could play an important role in the process of sugar metabolism in embryo cells [37].

H. OTHER ENZYMES

Cholinesterase activity was found in the water soluble fraction of the yolk by Willems and Stockx (1972) using acetylthiocholine iodide as a substrate [38]. In 1980, Saeed and his coworkers purified two cholinesterases, separating them into 1 and 2 by molecular sieving after delipidation followed by the chromatographic methods, to 12,100- and 230-fold, respectively [39]. From 2.3 l of yolk/water=1:1 mixture, peak 1 and peak 2 were recovered at 188 mU and 8.6 mU, respectively, determining the activity on acetylthiocholine as a substrate. The isoelectric points of the two isozymes are both 4.80, and the optimal pHs are both 8.0. Citrate and low concentrations of choline chloride activate the enzyme activity, whereas alkaloid eserine, a general cholinesterase inhibitor, blocks it. The enzymes are thought to be a glycoprotein because of the adsorption onto a Con A column.

α-Amylase exists in yolk and egg white and has been studied for stage dependence of activity after fertilization [1, 40].

The occurrence of catalase in hen eggs has been reported in which the residual activity of catalase after heat treatment at 50-60°C was suggested as an index of the degree of heat pasteurization of egg white [21].

III. DISCUSSION

As overviewed above, about 30 kinds of enzyme activities in unfertilized hen eggs have been detected so far. Several enzymes have been purified to homogeneity, and the localization and properties of the enzymes were determined. However, except for several excellent studies done in the recent few years, no particular progress has been reported on the enzymes in unfertilized hen eggs. Several of the the following reasons for this circumstance may be taken into consideration: 1) The low level of enzyme activities of hen eggs other than lysozyme makes the basic and applied study of egg enzymes difficult. 2) As scientific research materials, other creatures, such as *Drosophila*, nematodes, and amphibia, take more advantage of biochemical and molecular biological studies in morphogenesis, cell differentiation, and so on than chicken, because of easy handling, availability of mutant preparation, and short life cycle. 3) Most enzymes found so far in hen eggs are categorized as metabolic enzymes. Not only in egg yolk metabolism during embryogenesis but, in normal cells, the studies for *in vivo* degradation processes appear to be content with minor subjects. Nevertheless there are many fascinating subjects about enzymes of hen eggs. For example, the problems of whether an egg enzyme is associated with other compounds or not, how several enzymes accumulate in the eggs, which compounds are actual substrates for those enzymes, and whether some functional site(s) other than catalytic sites exists in those enzyme proteins having rather high molecular weight or not, are of interest and to be solved for elucidation of the biological significance of hen egg enzymes.

REFERENCES

1. **Lineweaver, H., Morris, H. J., Kline, L., and Bean, R. S.**, Enzymes of fresh hen eggs, *Arch. Biochem. Biophys.*, **16**, 443, 1948.
2. **Burley, R. W. and Vadehra, D. V.**, *The Avian Egg. Chemistry and Biology,* Burley, R. W. and Vadehra, D. V., Eds., John Wiley & Sons, New York, 1989, p 393.
3. **Wouters, J. and Stockx, J.**, Acid proteases from hen's egg yolk: Comparative investigation

of four enzyme species, purified through affinity chromatography on pepstatin-Sepharose, *Arch. Int. Physiol. Biochim.*, **89**, B211, 1981.
4. **Wouters, J., Goethals, M., and Stockx, J.**, Acid proteases from the yolk and the yolk-sac of the hen's egg. Purification, properties and identification as cathepsin D, *Int. J. Biochem.*, **17**, 405, 1985.
5. **Ichishima, E., Yamagata, Y., Chiba, H., Sawaguchi, K., and Tanaka, T.**, Soluble and bound forms of aminopeptidase in hen's egg yolk, *Agric. Biol. Chem.*, **53**, 1867, 1989.
6. **Petrovic, S. and Vitale, L.**, Purification and properties of glutamyl aminopeptidase from chicken egg-white, *Comp. Biochem. Physiol.*, **95B**, 589, 1990.
7. **Tanaka, T., Oshida, K., and Ichishima, E.**, Electron microscopic analysis of dimeric form of aminopeptidase Ey from hen's egg yolk, *Agric. Biol. Chem.*, **55**, 2179, 1991.
8. **Tanaka, T. and Ichishima, E.**, Molecular properties of aminopeptidase Ey as a zinc-metalloenzyme, *Int. J. Biochem.*, **25**, 1681, 1993.
9. **Tanaka, T. and Ichishima, E.**, Substrate specificity of aminopeptidase Ey from hen's (*Gallus domesticus*) egg yolk, *Comp. Biochem. Physiol.*, **105B**, 105, 1993.
10. **Tanaka, T. and Ichishima, E.**, Inactivation of chemotactic peptides by aminopeptidase Ey from hen's (*Gallus gallus domesticus*) egg yolk, *Comp. Biochem. Physiol.*, **107B**, 533, 1994.
11. **Skrtic, I. and Vitale, L.**, Methionine-preferring broad specificity aminopeptidase from chicken egg-white, *Comp. Biochem. Physiol.*, **107B**, 471, 1994.
12. **Moors, A. and Stockx, J.**, Phosphoesterase activities in the hen's egg I. Purification and characterization, *Arch. Int. Physiol. Biochim.*, **80**, 717, 1972.
13. **Moors, A. and Stockx, J.**, Phosphoesterase activities in the hen's egg. II. Phosphomonoesterases, *Arch. Int. Physiol. Biochim.*, **80**, 923, 1972.
14. **Debruyne, I.**, Hen's egg yolk acid phosphatase: Purification by hydrophobic chromatography; General characterization and kinetic properties, *Int. J. Biochem.*, **15**, 417, 1983.
15. **De Boeck, S. and Stockx, J.**, The acid phosphomonoesterase of the hen's egg vitelline membrane, *Arch. Int. Physiol. Biochim.*, **88**, B22, 1979.
16. **De Boeck, S. and Stockx, J.**, The alkaline phosphomonoesterase of hen's ovovitelline membrane, *Arch. Int. Physiol. Biochim.*, **86**, 935, 1978.
17. **Debruyne, I. and Stockx, J.**, Alkaline phosphatases from hen's egg yolk, liver, blood plasma and intestinal mucosa: Affinity chromatographic purification and comparison, *Int. J. Biochem.*, **10**, 981, 1979.
18. **Debruyne, I.**, Hen's egg yolk alkaline phosphatase: general characterization and kinetic study with inhibitors, *Int. J. Biochem.*, **14**, 519, 1982.
19. **Debruyne, I.**, Transphosphorylation mechanism of hen's egg yolk alkaline phosphatase, *Int. J. Biochem.*, **14**, 529, 1982.
20. **Lush, I. E. and Conchie, J.**, Glycosidases in the egg albumen of the hen, the turkey and the Japanese quail, *Biochim. Biophys. Acta*, **130**, 81, 1966.
21. **Henderson, A. E. and Robinson, D. S.**, Effect of heat pasteurisation on some egg white enzymes, *J. Sci. Fd Agric.*, **20**, 755, 1969.
22. **Donovan, J. W. and Hansen, L. U.**, The b-N-acetylglucosaminidase activity of egg white. 2. Heat inactivation of the enzyme in egg white and whole egg, *J. Food Sci.*, **36**, 174, 1971.
23. **De Boeck, S. and Stockx, J.**, a-Mannosidase in the hen's egg yolk, *Arch. Int. Physiol. Biochim.*, **82**, 976, 1974.
24. **Haaland, J. E. and Rosenberg, M. D.**, Activation of membrane associated ATPase in hen's egg after ovulation, *Nature*, **223**, 1275, 1969.

25. **Haaland, J. E., Etheredge, E., and Rosenberg, M. D.**, Isolation of an ATPase from the membrane complex of the hen's egg, *Biochim. Biophys. Acta,* **233**, 137, 1971.
26. **Etheredge, E., Haaland, J. E., and Rosenberg, M. D.**, The functional properties of ATPases bound to and solubilized from the membrane complex of the hen's egg, *Biochim. Biophys. Acta*, **233**, 145, 1971.
27. **Debruyne, I. and Stockx, J.**, Nucleoside triphosphatase from the hen's egg white and vitelline membrane. Purification, properties and relation to similar enzymes from the oviduct, *Enzyme*, **23**, 361, 1978.
28. **Willems, J., Teuchy, H., and Stockx, J.**, Fractionation, structure and some properties of yolk granules from unfertilized hen eggs, *Cytobios*, **17**, 195, 1976.
29. **De Moor, M. and Stockx, J.**, Aggregation-dissociation equilibria of the hen's egg white ribonuclease, *Arch. Int. Physiol. Biochim.*, **77**, 547, 1969.
30. **Wouters, J. and Stockx, J.**, A purification scheme for ribonuclease activity(ies) in hen's-egg yolk, *Biochem. Soc. Trans.*, **5**, 1104, 1977.
31. **Wouters, J. and Stockx, J.**, Comparative investigation of two ribonuclease species from hen's egg yolk, *Arch. Int. Physiol. Biochim.*, **85**, 1037, 1977.
32. **Wouters, J. and Stockx, J.**, Substrate specificity of ribonuclease from hen's egg yolk and albumen, *Arch. Int. Physiol. Biochim.*, **86**, 474, 1978.
33. **De Boeck, S. and Stockx, J.**, Nucleosidase and deaminase activities in the hen's egg, *Arch. Int. Physiol. Biochim.*, **80**, 957, 1972.
34. **De Boeck, S., Rymen, T., and Stockx, J.**, Adenosine deaminase in chicken-egg yolk and its relation to homologous enzymes in liver and plasma of the adult hen, *Eur. J. Biochem.*, **52**, 191, 1975.
35. **De Boeck, S. and Stockx, J.**, A purine N_1-C_6 hydrolase activity in the chicken egg yolk: A vestigial enzyme?, *Enzyme*, **23**, 56, 1978.
36. **Noda, S. and Schoner, W.**, Demonstration of a heterogeneous distribution of glycolytic enzymes and of pyruvate kinase isoenzymes types M_1 and M_2 in unfertilized hen eggs, *Biochim. Biophys. Acta*, **884**, 395, 1986.
37. **Noda, S., Horn, F., Linder, D., and Schoner, W.**, Purified pyruvate kinases type M_2 from unfertilized hen's egg are substrates of protein kinase C, *Eur. J. Biochem.*, **155**, 643, 1986.
38. **Willems, J. and Stockx, J.**, Structure bound enzymes in the yolk of the unfertilized hens egg, *Arch. Int. Physiol. Biochim.*, **80**, 989, 1972.
39. **Saeed, A., De Boeck, S., Debruyne, I., Wouters, J., and Stockx, J.**, Hen's egg yolk cholinesterase. Purification, characterization and comparison with hen's liver and blood plasma cholinesterase, *Biochim. Biophys. Acta*, **614**, 389, 1980.
40. **Ikeno, T. and Ikeno, K.**, Amylase activity increases in the yolk of fertilized eggs during incubation in chickens, *Poult. Sci.*, **70**, 2176, 1991.
41. **Ball, H. R. J. and Winn, S. E.**, b-*N*-acetylglucosaminidase activity of high pH egg white, *Poult. Sci.*, **50**, 1549, 1971.
42. **Donovan, J. W. and Hansen, L. U.**, The b-N-acetylglucosaminidase activity of egg white. 2. Heat inactivation of the enzyme in egg white and whole egg, *J. Food Sci.*, **36**, 174, 1971.
43. **Winn, S. E. and Ball, H. R., Jr.**, b-N-acetylglucosaminidase activity of the albumen layers and membranes of the chicken's egg, *Poult. Sci.*, **54**, 799, 1975.
44. **Moors, A. and Stockx, J.**, Alkaline and acid phosphomonoesterase and phosphodiesterase activities in chicken eggs. I., *Arch. Int. Physiol. Biochim.*, **74**, 728, 1966.
45. **Moors, A. and Stockx, J.**, Alkaline and acid phosphomonoesterase and phosphodiesterase activities in chicken eggs. II., *Arch. Int. Physiol. Biochim.*, **76**, 195, 1968.

46. **Debruyne, I. and Stockx, J.**, The nucleoside triphosphate and nucleoside diphosphate phosphohydrolase, the 5'-nucleotidase and alkaline phosphatase activities in the hen's egg yolk, *Arch. Int. Physiol. Biochim.*, **84**, 148, 1975.
47. **Debruyne, I. and Stockx, J.**, Some properties of the alkaline phosphatase from the hen's egg yolk, *Arch. Int. Physiol. Biochim.*, **85**, 961, 1977.
48. **Debruyne, I. and Stockx, J.**, Further evidence for the aspecificity of the alkaline phosphohydrolase from hen's egg yolk, *Biochem. Soc. Trans.*, **5**, 1109, 1977.
49. **De Boeck, S. and Stockx, J.**, A specific 5'-nucleotidase in the hen's egg vitelline membrane, *Arch. Int. Physiol. Biochim.*, **88**, B128, 1980.
50. **Jackle, M. and Geiges, O.**, Determination of alpha-amylase activity in whole egg and egg yolk by means of the phadebas^R amylase test (in German), *Mitt. Gebiete Lebensm. Hyg.*, **77**, 420, 1986.
51. **De Boeck, S. and Stockx, J.**, Nucleosidase and deaminase activities in the hen's egg, *Arch. Int. Physiol. Biochim.*, **82**, 957, 1972.
52. **De Boeck, S., Willems, J., Saeed, A. and Stockx, J.**, Cholinesterase activities in the hen's egg, *Arch. Int. Physiol. Biochim.*, **82**, 977, 1974.
53. **Fleming, A.**, On a remarkable bacteriolytic element found in tissues and secretions, *Roy. Soc. Proc., B.*, **93**, 306, 1922.
54. **Debruyne, I. and Stockx, J.**, Purification and ion-dependence of the nucleoside triphosphate phosphohydrolase of the hen's egg white, *Arch. Int. Physiol. Biochim.*, **81**, 963, 1973.
55. **Debruyne, I. and Stockx, J.**, Solubilization of the nucleoside triphosphate phosphohydrolase of the unfertilized hen's ovovitelline membrane, *Arch. Int. Physiol. Biochim.*, **83**, 175, 1975.
56. **Emanuelsson, H.**, Proteolytic activity in hen's egg prior to incubation, *Nature*, **168**, 958, 1951.
57. **Emanuelsson, H.**, Changes in the proteolytic enzymes of the yolk in the developing hen's egg, *Acta Physiol. Scand.*, **34**, 124, 1955.
58. **De Moor, M. and Stockx, J.**, Substrate specificity of ribonuclease from hen's egg yolk and albumen, *Arch. Int. Physiol. Biochim.*, **79**, 630, 1971.

Chapter 10

CELL PROLIFERATION-PROMOTING ACTIVITIES IN UNFERTILIZED EGGS

A. Seko and L.R. Juneja

TABLE OF CONTENTS

I. Introduction
II. Cell Proliferation-promoting Activities in Unfertilized Hen Eggs
    A. Cell Proliferation-promoting Activities in Egg Yolk
    B. Cell Proliferation-promoting Activities in Egg White
III. Discussion

References

I. INTRODUCTION

The culture cell system is one of the most available methods which helps to analyze biological phenomena *in vitro* involving so many complex factors. The system is an indispensable technique which offers, for example, the preparation of monoclonal antibodies with the aid of cell fusion methods, and the production of human-applicable physiologically active substances which can not be obtained from other sources. The addition of serum derived from domestic animal fetuses into a culture medium is very important for the maintenance, and proliferation of culture cells and the stable production of specific substances. All constituents of serum are, however, not clear, and the incomplete understanding sometimes causes complications in various biochemical analyses and purification processes. These obstacles have triggered the exploration of serum-free media by replacement of serum with a variety of well-characterized growth-promoting factors or cell adhesion proteins. Hen egg substances have been studied as one of the candidates for a substitute because of their availability, safety, and cheapness. Another reason for the trial of hen eggs is that there could be certain intrinsic substances promoting cell growth or differentiation in hen eggs during embryogenesis, even in unfertilized eggs, because the events occur naturally and dynamically during developmental stages. Hen egg substances used for various cell cultures studied so far are shown in Table 1. It is obvious that hen egg substances possess cell growth-promoting activities for various cell lines. It should be noted, however, that many cell lines need other well-known factors, such as transferrin, ethanolamine, or sodium selenite, which do not stimulate cell growth by themselves, but support practical applications of the various stimulating effects of hen egg components. In this chapter, the cell proliferation-promoting activities residing in unfertilized eggs are overviewed.

II. CELL PROLIFERATION-PROMOTING ACTIVITIES IN UNFERTILIZED HEN EGGS

In early experiments, extracts of 10- and 16-day-embryos, in which cell proliferation occurs vigorously, had been used but unfertilized eggs were recognized as a more convenient substitute. In the 1950's Earle and his colleagues [1] and Evans and his colleagues [2] reported that ultrafiltrated extracts of whole eggs, 10- day-, or 16-day-embryos have growth-promoting activity for the mouse fibroblast L-929, mouse liver 1469, and the human cervical carcinoma HeLa strain, although the determination of the cell growth-promoting effect was still not quantitative because of incompleteness in the method of measuring cell growth. Walker (1967) studied the growth and survival rate of chick embryos, which were displaced surgically from the yolk fluid, with various yolk fractions derived from unfertilized eggs [3]. As a result, floating fractions rich in lipoprotein prepared by centrifugation supported the growth of embryos, but the survival rate for the fraction was a little lower than the control given by whole yolk. Conversely, a sedimented granule fraction showed a survival rate equal to the control while the growth of the embryos was at a low level. Thus, Walker offered an interesting suggestion that the granule fraction might contain some compound, probably phosphoproteins, supporting the survival of chick embryos whereas the floating fraction might have some embryonic growth-promoting factors. Thereafter, the active substances in hen eggs were divided into two streams to use as a source for yolk lipoprotein and egg white.

Table 1

Hen Egg Components Promoting Cell Growth or Injuring Tumor Cells

Component	Cell line	Reference
Whole egg extract	L-929 cells	1, 2
	mouse liver 1469 cells	1, 2
	HeLa cells	1, 2
Yolk	adult bovine aortic endothelium	5
	bovine corneal endothelial cells	5
	vascular smooth muscle cells	5
	avian tendon cells	6
	equine embryos	13
Yolk lipoprotein	chicken embryos	3
	human B cell lines (Bri-7, HMy-2, RPMI-6666, Oda, K103, Wil-2, K562)	7
	human T cell lines (HSB-2, CEM, Molt-4)	7
	human B cell hybridoma (HB4C5)	7, 10, 12
	human lung adenocarcinoma (PC-8)	7
	human melanoma (Bowes)	7
	human breast adenocarcinoma (ZR75-1)	7
	human stomach adenocarcinoma (MKN-28)	7
	mouse myelomas (P3U1, X63-653)	7
	human B cell hybridomas (HB4C5, HF10B4, K-1-5, H15F1, 9P-13-4)	9
	mouse B cell hybridoma (D6, F6)	9
	human lymphoma cells (U-937)	11
	human monocytic leukemia cell line (THP-1)	11
	macrophage-like cell line (U-M)	11
	human-human hybridomas	11
Yolk low density lipoprotein	human hepatoma (Hep G2)	8
Yolk very low density lipoprotein	mouse-human hybridoma (4H11)	21
Phosphatidylcholine	adult bovine aortic endothelial cells	4
	bovine corneal endothelial cells	4
	vascular smooth muscle cells	4
Egg white	kidney fibroblast cells (BHK-21, CV-1)	17
	rat epithelial cells (IEC-18)	17
	rat adrenal pheochromocytoma (PC12)	17
Acid extract of egg white	BALB/c 3T3 cells	16
Conalbumin	human peripheral lymphocytes	18
Ovomucin	human melanoma (SEKI cell)	20
	mouse Lewis lung cancer-derived 3LL cells	20
	mouse spleen lymphocytes	19

A. CELL PROLIFERATION-PROMOTING ACTIVITIES IN EGG YOLK

Fujii and his colleagues (1983) found that human high density lipoprotein (HDL), which can stimulate the growth of adult bovine aortic endothelium cells (ABAE), bovine corneal endothelial cells (BCE), and vascular smooth muscle cells (VSM), can be replaced in these bioassay systems by the liposome of phosphatidylcholine (PC) which is one of the major components of human HDL [4]. In these systems, phosphatidylethanolamine, sphingomyelin, phosphatidylserine and phosphatidylglycerol were not available like PC. Moreover, the effect of PC liposome on these cell growth-promoting activities was found to be significantly dependent on its fatty, acid composition. PC derived from egg yolk has stronger activity than that derived from plants. The fact that PC is the major phospholipid in egg yolk had led Fujii and Gospodarowicz to examine the effect of egg yolk on the growth of ABAE, BCE, and VSM cells on the plate coating basement lamina; it had turned out that egg yolk has a level of activity almost equal to bovine calf serum [5].

Martis and Schwarz (1986) has found that the yolk supernatant fraction, although derived from the fertilized eggs, promotes the growth of primary chick tendon cells and the biosynthesis of procollagen and that the fraction contains four major proteins, 82 K, 70 K, 42 K, and 35 K as proteinaceous compounds [6]. The proteinaceous components were suggested by them to play an important role in the promoting activity of the fraction because of the sensitivity of the activity to heat or proteinase treatment.

Murakami and his colleagues (1988) reported that yolk lipoprotein fraction prepared by gel filtration and isoelectric focusing (around pI 7.5) stimulates the growth of mouse myeloma line X63-653 [7]. The activity of this fraction, which seemed to be very low density lipoproteins because of its high lipid content (96.1 %), was extinguished when the fraction was divided into a lipid fraction and residual apoproteins, neither of which shows activity by itself; it had been indicated that the form of protein-lipid complex is important for the promoting activity. Nakama and Yamada (1993) have also shown that yolk LDL is effective in the growth of human hepatoma Hep G2 cells [8].

Recently, large-scale production of human-type monoclonal antibodies by human-human hybridoma cells has been studied for application as refined medicines. In this field, too, well characterized and safe factors able to stimulate cell growth and to enhance the secretion of immunoglobulins are needed. Egg yolk lipoproteins have offered good results for such demands. Yamada and his colleagues (1989) reported that yolk lipoprotein can stimulate the production of immunoglobulins by human-human hybridoma cells in some cases; yolk lipoprotein is effective, in general, in the production of the IgM class, but it is ineffective in the IgG class [9]. Shinohara and his colleagues (1993) showed that the γ-livetin- and HDL-rich fractions have IgM secretion-promoting effects, though they are ineffective in promotion of the growth of human-human hybridoma cells (HB4C5) [10-12]. Interestingly, PC was proved to have effects similar to that fraction.

Egg yolk was shown to have the effect of supporting mammal embryogenesis. Equine embryos grew normally with high ratios in the presence of 20-60 % of relatively high content yolk [13], although it was ineffective for those of the murine [14] and bovine embryos [15].

The optimal concentration of yolk lipoproteins for the cell growth-promoting effect was found to be approximately 100-500 μg/ml by several independent studies using different types of cell lines. The essentially active component of yolk lipoproteins is due to PC alone in some cell lines, whereas it is a protein-lipid complex in others; the discrepancy might be attributed to the difference between the sensitivities of those cell lines. The liposome form of PC examined in the study of Fujii and his colleagues (1983) might be important for its cell growth-promoting activity [4]. Since yolk lipoprotein is easily prepared by centrifugation or ammonium sulfate fractionation followed by anion-exchange chromatography and easily separable from culture medium, it will increasingly be applied to serum-free media.

B. CELL PROLIFERATION-PROMOTING ACTIVITIES IN EGG WHITE

Azuma and his colleagues (1989) reported on egg white peptidic factors effective in culture cells, such as BALB/c 3T3 cell growth [16]. Two fractions effective in promoting DNA synthesis of the cells were prepared by 1 N AcOH/1 N HCl/1 M NaCl extraction followed by Sephadex G-25 gel filtration. The molecular weights of the two fractions were about 7 K and 2 K. These fractions include peptidic growth promoting factors because their activity is sensitive to heat or "Pronase" digestion. They suggested that the 7 K fraction, sensitive to SH-reducing reagent, seems to contain an EGF-like factor.

Zou and his colleagues (1991) have observed the cell growth- or neurite growth-promoting effects on mammalian cells [17]. The culture medium containing 10% egg white promotes growth, even if to a lesser extent than fetal bovine serum, of kidney fibroblast cell CV-1. The cell spreading-promoting activity of egg white is, however, present at levels equal to serum and the outgrowth-promoting activity is very strong whereas serum shows no such effects. Egg white was also able to induce neurite outgrowth for rat adrenal pheochromocytoma PC 12 cells. The authors suggested that, since the neurite outgrowth of PC 12 cells is also supported by nerve growth factor (NGF), egg white might contain an NGF-like substance or an adhesion molecule like laminin and fibronectin, which are also capable of supporting PC 12 cell spreading and neurite outgrowth.

In 1983 Mantovani and his colleagues shown that conalbumin can enhance the stimulating effects of phytohemagglutinin or pokeweed mitogen on human peripheral lymphocytes [18]. Otani and Maenishi reported that an ovalbumin- and avidin-rich fraction suppresses the growth of mouse spleen lymphocytes, whereas an ovomucin-rich fraction promotes the growth of the lymphocytes activated by a lipopolysaccharide derived from *Salmonella typhimurium* [19]. Ovomucin has also been shown to have tumor cell-injuring activity. Ohami and his colleagues (1993) prepared a sialic acid rich fraction from ovomucin and found that, with human malignant melanoma (SEKI cell) and mouse Lewis lung cancer-derived 3LL cells, the fraction causes morphological changes on the cell surface and a decrease of living cell numbers [20]. They suggested that ovomucin attacks the cell surface of tumor cells.

III. DISCUSSION

Since biochemical information on the various factors affecting cell growth mentioned is being explored and, in some cases, the substances possess the cell growth-promoting activity equal to serum, basic techniques for industrial applications are being established.

On the other hand, scientific studies of those active substances involved in molecular action on culture cells and with biological significance in unfertilized hen eggs and during embryogenesis are a long way from being investigated. Since it is evident that developmental processes proceed with strict regulation for both cell growth and cell differentiation, it may be suggested that the localization and physiological activity of those substances are exactly controlled if present during embryogenesis. It should be considered whether these activities could be modified by proteinases and lipases expressed during the degradation process of yolk and egg white or not, since those activities, so far found, are proteinaceous or phospholipids by nature.

Anyway, it is clear that egg yolk lipoproteins and some compounds of the egg white are effective in the cell growth of many cell lines. Since egg substances have the potential to bring on embryogenesis themselves, it could be found that hen eggs contain several activities to regulate various cell responses besides cell growth-promoting activities.

REFERENCES

1. **Earle, W. R., Bryant, J. C., Schilling, E. L., and Evans, V. J.,** Growth of cell suspensions in tissue culture, *Ann. N. Y. Acad. Sci.*, 63, 666, 1956.
2. **Evans, V. J., Bryant, J. C., McQuilkin, W. T., Fioramonti, M. C., Sanford, K. K., Westfall, B. B., and Earle, W. R.,** Studies of nutrient media for tissue cells *in vitro* III. Whole egg extract ultrafiltrate for long-term cultivation of strain L cells, *Cancer Res.*, 17, 317, 1957.
3. **Walker, N. E.,** Growth and development of chick embryos nourished by fractions of yolk, *J. Nutr.*, 92, 111, 1967.
4. **Fujii, D. K., Cheng, J., and Gospodarowicz, D.,** Phosphatidylcholine and the growth in serum-free medium of vascular endothelial and smooth muscle cells, and corneal endothelial cells, *J. Cell. Physiol.*, 114, 267, 1983.
5. **Fujii, D. K. and Gospodarowicz, D.,** Chicken egg yolk-supplemented medium and the serum-free growth of normal mammalian cells, *In Vitro*, 19, 811, 1983.
6. **Martis, M. J. and Schwarz, R. I.,** A simple fractionation of chicken egg yolk yields a protein component that stimulates cell proliferation and differentiation in primary avian tendon cells, *In Vitro Cell. Dev. Biol.*, 22, 241, 1986.
7. **Murakami, H., Okazaki, Y., Yamada, K., and Omura, H.,** Egg yolk lipoprotein, a new supplement for the growth of mammalian cells in serum-free medium, *Cytotechnology*, 1, 159, 1988.
8. **Nakama, A. and Yamada, A.,** Serum-free culture of human hepatoma Hep G2 cells with egg yolk low density lipoprotein, *Biosci. Biotech. Biochem.*, 57, 410, 1993.
9. **Yamada, K., Ikeda, I., Sugahara, T., Shirahata, S., and Murakami, H.,** Screening of immunoglobulin production stimulating factor (IPSF) in foodstuffs using human-human hybridoma HB4C5 cells, *Agric. Biol. Chem.*, 53, 2987, 1989.
10. **Shinohara, K., Kakiuchi, T., Kobori, M., Suzuki, M., and Fukushima, T.,** Europium-linked immunoassay of IgM production by human-human hybridomas cultured with or without chicken egg-yolk lipoprotein, *Biosci. Biotech. Biochem.*, 57, 1012, 1993.
11. **Suzuki, M., Shinmoto, H., Yonekura, M., Tsutsumi, M., and Shinohara, K.,** Growth promoting effect of chicken egg yolk lipoproteins on human lymphocytic cell lines, *Nippon Shokuhin Kogyo Gakkaishi (in Japanese)*, 41, 37, 1994.
12. **Shinohara, K., Fukushima, T., Suzuki, M., Tsutsumi, M., Kobori, M., and Kong, Z.-L.,** Effect of some constituents of chicken egg yolk lipoprotein on the growth and IgM production of human-human hybridoma cells and other human-derived cells, *Cytotechnology*, 11, 149, 1993.
13. **Lebedev, S. G. and Lebedeva, L. F.,** Culture of equine embryos in media containing egg yolk, mare's milk and saline: Preliminary results, *Theriogenology*, 41, 1201, 1994.
14. **Hammond, J.,** Recovery and culture of tubal mouse ova, *Nature*, 263, 28, 1949.
15. **Dowling, D. F.,** Problems of the transplantation of fertilized ova, *J. Agric. Sci.*, 39, 374, 1949.
16. **Azuma, N., Mori, H., Kaminogawa, S., and Yamauchi, K.,** Presence of cell growth promoting peptides in chicken egg white extract, *Agric. Biol. Chem.*, 53, 1291, 1989.
17. **Zou, C., Kobayashi, K., and Kato, A.,** Effects of chicken egg white on the proliferation and neurite outgrowth of mammalian cells, *J. Agric. Food Chem.*, 39, 2137, 1991.
18. **Mantovani, G., Puddu, A., Leone, A. L., Tognella, S., and Del Giacco, G. S.,** Effect of conalbumin on phytomitogen stimulation and E-rosette formation of human peripheral lymphocytes in normal subjects, *Int. J. Tiss. Reac.*, V, 107, 1983.
19. **Otani, H. and Maenishi, K.,** Effects of hen egg proteins on proliferative responses of

mouse spleen lymphocytes, *Lebensm. Wiss. u. Technol.*, **27**, 42, 1994.
20. **Ohami, H., Ohnishi, H., Yokota, T., Mori, T., and Watanabe, K.,** Cytotoxic effect of sialoglycoprotein derived from avian egg white ovomucin on the cultured tumor cell, *Med. Biol. (in Japanese)*, **126**, 19, 1993.
21. **Takazawa, Y., Tokashiki, M., Murakami, H., Yamada, K., and Omura, H.,** High-density culture of mouse-human hybridoma in serum-free defined medium, *Biotechnol. Bioeng.*, **31**, 168, 1988.

Chapter 11

EGG YOLK ANTIBODY IgY AND ITS APPLICATION

H. Hatta, M. Ozeki, and K. Tsuda

TABLE OF CONTENTS

I. Introduction
II. Egg Yolk as a Source of Antibodies
    A. Antibodies and Humoral Immunity
    B. Passive Immunity of Hens
    C. Feature of Using Hen Eggs for Preparation of Antibodies
    D. Productivity of IgY
        1. Quantity of Antibodies Obtained
        2. Antibodies Specific to Rotavirus
        3. Antibodies Specific to Mouse IgG
        4. Antibodies Specific to Human Insulin
III. Methods for Purification of IgY
    A. Strategy of Purification of IgY
    B. Purification of IgY Using λ-carrageenan
    C. Some Properties of Purified IgY
IV. Some Differences in Properties Between IgG and IgY
    A. Structure and Immunological Functions
    B. Heat and pH Stability
    C. Stability to Several Proteolytic Enzymes
V. Application of IgY as an Immunological Tool
    A. Diagnosis
    B. IgY as a Ligand of an Immunoadsorbent
VI. Passive Immunization by Use of IgY
    A. Systemic Administration of IgY
        1. Prevention of Newcastle Disease Virus
        2. Neutralization of Venom
    B. Oral Administration of IgY
        1. Prevention of Rotaviral Diarrhea
        2. Prevention of Fish Disease
        3. Prevention of Dental Caries
References

I. INTRODUCTION

The antigen-specific IgG isolated from sera or colostrum of superimmunized animals, such as rabbits, cows, and goats has been used widely as an immunological tool in the fields of diagnosis as well as pure scientific research. Another important application of IgG is for passive immunization therapy in which IgG specific to the antigen of a certain infectious disease (pathogen) is administered to unimmunized individuals for prevention of the disease. Several researchers have demonstrated that the effect of passive immunization is quite effective when IgG is administered systemically or orally to individuals for preventing infectious diseases. However, the practical application for passive immunization therapy remained questionable, because a large excess of IgG was supposed to be required for the therapeutic treatment.

The IgG found in blood serum of the hen is known to be transferred to the yolk of eggs laid by the hen to give acquired immunity to the offspring. The antibody in egg yolk has been referred to as IgY (Yolk Immunoglobulin) [1], because its protein nature is somewhat different from mammalian IgG in structure and immunological properties. At present, a tremendous number of hens are being systematically immunized with several antigens (vaccination) to protect the hens from infectious diseases and managed to lay eggs as scheduled for commercial production. Hen eggs, therefore, are now considered to be a potential source of large-scale production of antibody (IgY). Moreover, eggs containing IgY specific to Newcastle disease virus, as an example, which causes Newcastle disease in hens have been consumed as food for a long time. This indicates that there seems to be no problem with eggs containing IgY specific to given antigen as food from the safety stand point. The above fact, at the same time, may indicate that the IgY specific to a given antigen is possibly useful for oral administration for preventing the diseases due to the antigens. Furthermore, IgY, which can be induced and isolated much more easily than antibodies from mammals probably can be used not only as an immunological tool in the field of diagnosis, but also for immobilized antibody as a ligand in the field of immuno-affinity chromatography.

Figure 1 shows a difference in the preparative procedure of an antibody specific to a given antigen when rabbit or hen is used. It is now possible to produce IgY in large amounts through the eggs from the hens immunized with a given antigen (proteins, bacteria, and viruses) instead of producing IgG through the blood from a mammal.

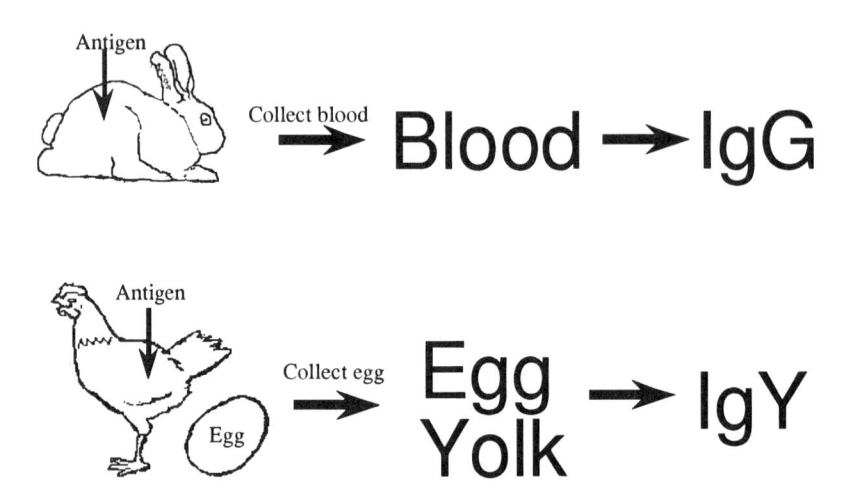

Figure 1. Illustrative preparation of a specific antibody.
Reproduced from Hatta et al. (1994) [52] by permission of CAB International, Oxon.

II. EGG YOLK AS A SOURCE OF ANTIBODIES

A. ANTIBODIES AND HUMORAL IMMUNITY

Antibodies sometimes have been referred as immunoglobulins (Igs), and are found in the humor (blood, saliva, milk, etc.) of mammals. Five classes of immunoglobulins (IgG, IgA,

IgM, IgD, and IgE) are known which are distinguishable in structure and immunological function. The major immunoglobulin is immunoglobulin G class (IgG) and it makes up about 75% of the Igs in circulating blood. Animals produce Igs against almost all kinds of antigens (bacteria, virus, and foreign proteins, etc.) in their humor to defend against invading substances. The Igs thus produced bind their antigens specifically to neutralize their effects. This biological defense system is a humoral immunity established evolutionarily in animals.

B. PASSIVE IMMUNITY OF HENS

In chickens, there have been found three kinds of Immunoglobulins (IgG, IgA, and IgM) in their circulating blood, and their concentrations have been reported to be 5.0, 1.25, and 0.61 mg for IgG, IgA, and IgM per 1 ml serum, respectively [2]. The hen transfers the immunoglobulins to her eggs in which IgM and IgA are in egg white at concentrations of about 0.15 mg and 0.7 mg per ml, respectively. On the other hand, IgG locates only in egg yolk in a concentration of about 25 mg per ml [3]. Hen's serum IgA and IgM are secreted together with other proteins to be the component of egg white at the oviduct, whereas the serum IgG is specifically transferred through a membrane into the yolk during its maturation (Figure 2). A receptor specific to IgG translocation is known to exist on the surface of the yolk membrane.

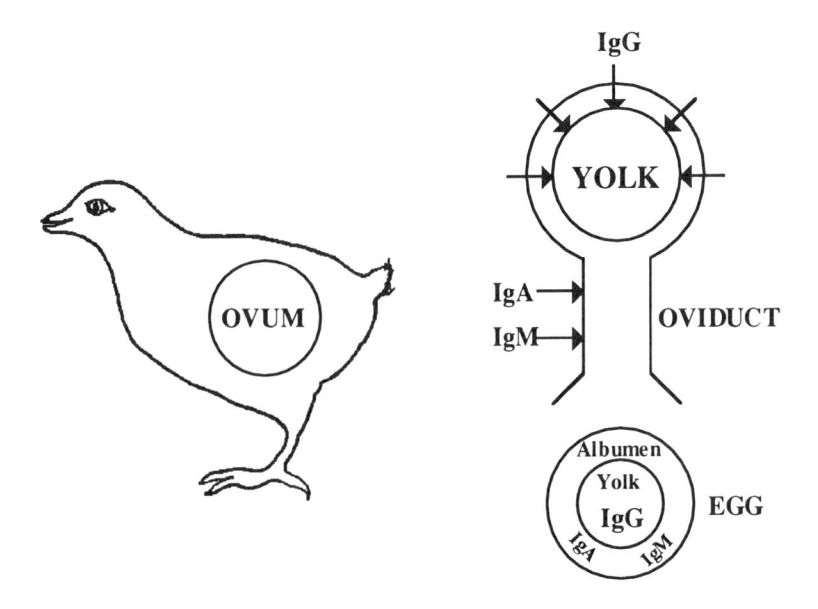

Figure 2. Distribution of immunoglobulins in hen egg.

It has been suggested that Igs in eggs are passive immunity, in a sense, because the antibodies in eggs originated from the mother hen are used to protect the new offspring from various infectious diseases. In fact, IgG found in egg yolk circulates in the blood and IgA and IgM in the digestive tract of the hatched chick. The antibodies transferred from hen to chick via the latent stage of the egg has thus an important immunological meaning for the newly developed chicks to resist various infectious diseases. This system of transfer of antibodies to offspring via the biologically latent form of eggs in birds is considered to be the same as the transfer of maternal immunity from the mother to the fetus via her placenta in mammals. The transfer of hen's immunity to her chick via the egg was first reported by Klemperer about 100 years ago. Since his finding, many researchers have reported that hens immunized with certain several antigens (bacteria, virus, and proteins) produced the antibodies and transferred them to their egg yolk.

C. FEATURE OF USING HEN EGGS FOR PREPARATION OF ANTIBODIES

An antigen-specific IgG has been conventionally isolated from sera of animals, such as rabbits, which have been superimmunized with an aimed antigen. IgY is also able to be isolated from the egg yolk laid by the hen superimmunized previously. The preparation methods of IgG and IgY are shown comparatively in Figure 1. Several advantages in the preparation of antibody using hens over using animals are summarized as follows:

(1) The conventional method inevitably sacrifices animals which have produced the specific IgG in their circulating blood. On the other hand, the method of using hens is sufficient only to collect the eggs laid by the superimmunized hens. For separating IgY, a large scale method is now applicable by automatic separation of the egg yolk with a machine (Figure 3).

(2) As egg yolk contains only IgY, the isolation of IgY from the yolk is much easier than that of IgG from animal blood sera.

(3) Large-scale feeding of hens for egg production now being carried out is also a merit for collecting the source of a specific antibody.

(4) Also, immunization of hens (vaccination) has long been applied to prevent hens from infectious diseases, indicating that immunization of hens is much more systematized to be effective than doing it for animals (Figure 3).

(5) Egg yolk as the source of IgY is much more hygienic than mammal's sera from which IgG is separated.

(6) Because of a taxonomical difference, the hen has the possibility to produce the antibodies whose formation is difficult or impossible in mammals.

D. PRODUCTIVITY OF IgY

Several papers have been published on a comparative study of the productivity of antibodies through hen egg and rabbit serum. Jensenius and his co-workers estimated the total antibody activity of the eggs laid by a hen in a month to be equivalent to that produced in a half liter of serum from an immunized rabbit [4]. Gottstein and Hemmeler compared the antibody production efficiency for *Echinococcus granulosus* as an antigen [5]. They reported that the quantity of IgY obtained from eggs laid by an immunized hen was 18 times greater than that of IgG isolated from the serum of an immunized rabbit. The authors also compared the productivity of IgY from the eggs laid by a hen over a year with that of IgG from the whole serum of a rabbit in which both animals were immunized with the same several antigens. The results are summarized in Table 1.

1. Quantity of Antibodies Obtained

Hens usually lay about 250 eggs (about 4,000 g egg yolk) in a year. On the other hand, the serum to be isolated from a rabbit is only 40 ml or so. One gram of egg yolk laid by the immunized hen is estimated to contain about 10 mg of IgY whereas 1 ml of rabbit serum yields about 35 mg of IgG. Therefore, it follows that an immunized hen yields 40 g of IgY through her eggs a year. However, IgG to be produced by an immunized rabbit is only 1.4 g or so. In other words, the productivity of antibody of hens is nearly thirty times greater than that by rabbits based on the weight of antibody produced per head per year.

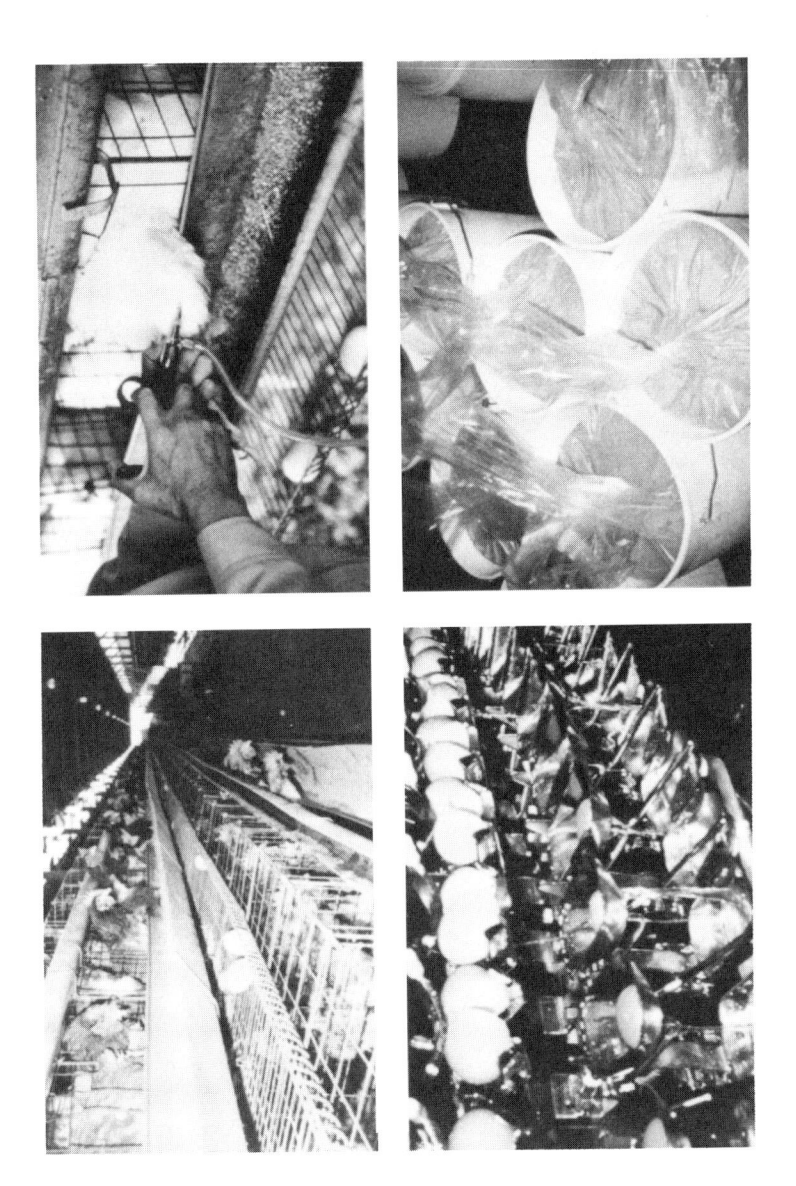

Figure 3. Production of eggs from immunized hens and collecting egg yolk from the eggs.

Table 1
Productivity of IgY and IgG

Immunized animal	Rabbit	Hen
Source of antibody	Blood serum	Egg yolk
Kind of antibody	Polyclonal	Polyclonal
Quantity of antibody	1,400 mg / rabbit	40,000 mg /hen
(Quantity of specific antibody)		
Anti-HRV (MO) antibody	5.6×10^6 NT	600×10^6 NT
Anti-HRV (Wa) antibody	37.8×10^6 NT	520×10^6 NT
Anti-Mouse IgG antibody	700 mg	11,200 mg
Anti-Insulin Antibody	0 mg	2,000 mg

NT: Neutralization titer
HRV: Human rotavirus

2. Antibodies Specific to Rotavirus

The productivity of IgY and IgG was compared on the basis of specific antibody activity toward human rotavirus (HRV) Wa and MO strains as the antigens. Hens were immunized four times intramuscularly with HRV using Freund's complete adjuvant at intervals of two weeks. Rabbits were also immunized with HRV in the same manner as done for hens and made to bleed one week after the fourth immunization. The antibody activity was performed by determining the neutralization titer. The neutralization titer of egg yolk against HRV increased drastically after the second immunization and its maximum was achieved immediately after the fourth immunization (Figure 4). The change of neutralization titer of rabbit serum was observed to be almost the same in the tendency as that of egg yolk. IgY was isolated from eggs laid between 8 to 12 weeks after the initial immunization. Rabbit IgG was isolated from the whole serum of the rabbit fed one week after the fourth immunization. Neutralization titers of anti-HRV IgY and IgG (1 mg/ml) against HRV are compared. The titer of anti-Wa IgY (13,000) against Wa strain was nearly half of that of anti-Wa IgG (27,000), while the titer of anti-MO IgY (15,000) against MO strain was four times as large as that of anti-MO IgG (4,000). The difference in the quantity and antigen neutralization titer between the two antibodies as described above will make it clear that the productivity of IgY by a hen, at least for anti-HRV IgY is 14 times (anti-Wa) or 110 times (anti-MO) greater than that of a rabbit immunized, as far as it is determined per head of animal (Table 1). Moreover, it should be noted that the production of antibody through hen eggs will be achieved without sacrificing or bleeding the animals, unlike through mammals.

3. Antibodies Specific to Mouse IgG

Mouse IgG was used as an antigen to compare the productivity of the antibody by hen and rabbit. The hens and rabbits were immunized four times intramuscularly with 1 mg of antigen using Freund's complete adjuvant at two week intervals. Antibody activities were determined

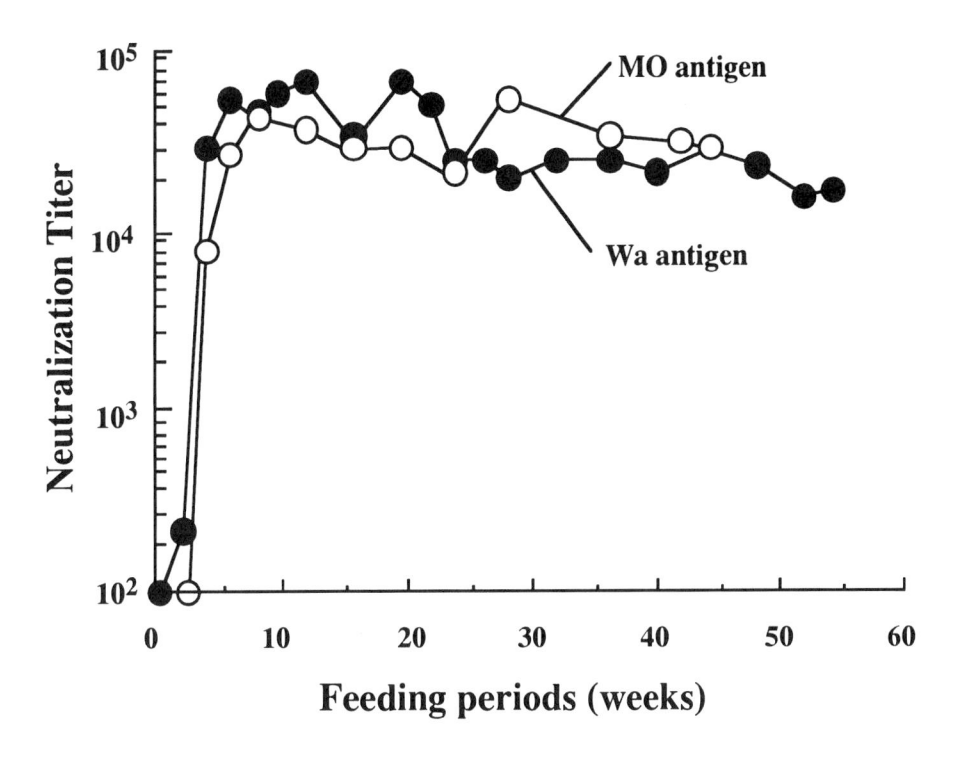

Figure 4. Change in neutralization titer against HRV Wa and MO of the yolk of eggs laid after immunization. Hens were immunized intramusculary at 0, 2, 4, 6, 22, and 38 weeks.

by ELISA. IgY activity against mouse IgG in egg yolk appeared and increased in one week after appearance of antibody activity in hen's serum (Figure 5). The IgY was purified from the eggs (8-24 weeks), and the rabbit IgG was purified from the whole rabbit serum obtained one week after the fourth immunization. Since both IgY and rabbit IgG thus obtained were polyclonal antibodies, the quantities produced of the antibody specific to mouse IgG were determined by using affinity-chromatography in which mouse IgG was conjugated with Sepharose 4B. The percentage of antibody bound to mouse IgG in IgY protein and that of IgG protein were 28% and 50%, respectively. As described above, an immunized hen usually yields 40 g of IgY protein through her eggs per year while IgG produced by an immunized rabbit is only 1.4 g. Therefore, the total amount of specific IgY (11,200mg) derived from a hen per year was estimated to be about 16 times as much as that of specific IgG (700mg) from a rabbit.

4. Antibodies Specific to Human Insulin

The productivity of antihuman insulin antibody was also compared using hens and rabbits as immunization animals. It is known that the amino acid (AA) sequence of insulin is highly preserved in mammals. Therefore, human insulin is generally less immunogenic in rabbits. The antihuman insulin antibody has thus been conventionally prepared by immunizing guinea pig whose insulin differs at 13 sites in the AA sequence from that of the human. The AA sequence of chicken insulin is also different from that of human insulin at 9 sites in the AA sequence. Therefore, it was presumed to be possible for the hen to produce antihuman insulin IgY. Our research group demonstrated that the hens superimmunized with human insulin produced polyclonal antibodies specific to human insulin which was identified by ELISA.

The IgY isolated from the eggs laid by the hens mentioned above, was found to contain about 5% of the IgY bound to human insulin-immobilized Sepharose 4B in affinity chromatography. On the other hand, antihuman insulin antibody was not produced in the rabbit.

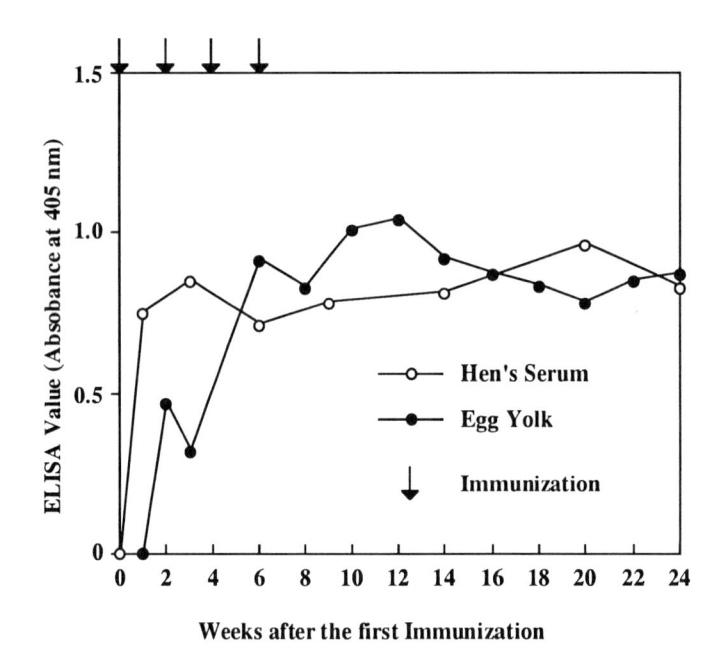

Figure 5. Change of IgY antibody activity in hen's serum and egg yolk against mouse IgG. Arrows indicate the weeks after immunization.

III. METHODS FOR PURIFICATION OF IGY

A. STRATEGY OF PURIFICATION OF IgY

Egg yolk is a fluid emulsion of a continuous phase of water-soluble protein (livetins) and a dispersed phase of lipoproteins. The immunoglobulin of egg yolk, now referred to as IgY, is known to be γ-livetin itself. Egg yolk contains two other livetins, α–livetin (chicken serum albumin) and β–livetin (α_2-glycoprotein) present together with lipoproteins (LDL and HDL) which are the major components of egg yolk. Therefore, the first step in purification of IgY is to separate the water-soluble proteins from lipoproteins (or lipids). Many purification methods of IgY have been reported so far: separation of lipoprotein by ultracentrifugation [6], delipidation by organic solvents [7], and precipitation of lipoproteins by polyethyleneglycol [8], sodium dextran sulfate [4], polyacryl acid resins [9], etc. These methods, however, were questionable to apply practically for isolation of IgY on a large scale. The above methods seemed to be costly and not necessarily effective for isolation of IgY. Also, the IgY prepared using organic solvents or chemicals for delipidation seemed to have some problems in safety.

Formerly, our group reported a simple method to separate water-soluble proteins in egg yolk using sodium alginate [10]. In the experiments successively carried out to find more effective gums for separation of egg yolk lipoproteins, carrageenan and xanthan gum were found to be highly effective as a precipitant of yolk lipoproteins [11]. The mechanism involved in precipitation of yolk lipoprotein by the gums has remained unsolved. However, the above effective gums are slightly acidic due to sulfonic or carboxyl groups, and thus there may be the possibility that the force to cause agglutination with lipoproteins is by ionic binding elaborated between the negative charge of the gums and the net positive charge of the lipoproteins. The effective gums selected above have been used as food ingredients, and thus their use for purification of IgY will present no problems, as far as the IgY thus purified is used for oral administration.

B. PURIFICATION OF IgY USING λ-CARRAGEENAN

Based on the findings mentioned above, a purification method for IgY was established using λ-carrageenan followed by column chromatography and salting-out with sodium sulfate (Figure 6) [11]. The egg yolk doubly diluted with water (200ml) was mixed with 400 ml of 0.15% λ-carrageenan solution. The mixture was left at room temperature (20°C) for 30 min followed by centrifugation at 10,000 x g for 15 min. The supernatant was filtered through a filter paper No. 2 (Advantic Toyo). Disodium hydrogen phosphate was added to this filtrate to be 20 mM, and the pH of the mixture was adjusted to 8.0. The mixture was then applied onto a DEAE-Sephacel column (φ 5.0 x 10.0 cm; Pharmacia) previously equilibrated with 20 mM phosphate buffer, pH 8.0, at a flow rate of 5.0 ml per min. After washing the column with the same buffer, the protein adsorbed was eluted using five column volumes of 200 mM phosphate buffer, pH 8.0. The peak fractions of eluted protein estimated by monitoring at 280 nm was added with anhydrous sodium sulfate to be 15% (w/v) at 20°C. The mixture was gently stirred for 30 min at 20°C and, then, centrifuged at 10,000 x g for 15 min. The resulting precipitate was dissolved in 100 ml of 100 mM phosphate buffer, pH 8.0. This procedure of salting-out with sodium sulfate was repeated three times. The precipitate finally obtained was dissolved in 10 mM of phosphate buffer, pH 8.0, and dialyzed against the same buffer. The supernatant of the dialyzed solution was filtered through a 0.45 μm membrane filter, and the filtrate was lyophilized. The purity of IgY thus obtained was 98.3% based on protein nature and its recovery yield 73%, as shown in Table 2.

Hassl and his co-workers reported a comparative study of purification methods for IgY [12]. They obtained the best result by using polyethyleneglycol (PEG 6,000), but the purity of IgY and its yield were 85% and about 40 mg per egg, respectively. In our method of using λ-carrageenan to remove lipoprotein at the first step of purification, the recovery yield of IgY was increased about 70% (70-100 mg per egg) with a purity of 98%.

Table 2

Summary of Purification of IgY from Egg Yolk

	Total protein Quantity (mg)	Yield (%)	Total lipids Quantity (mg)	Yield (%)	Total IgY Quantity (mg)	Yield (%)	Purity of IgY(b) (%)
Doubly diluted yolk (a)	16,000	100	29,000	100	790	100	4.9
Supernatant	3,500	22	140	0.5	680	86	19.4
Eluate	1,300	8.1	0	0	600	76	46.2
Salting-out solution	590	3.7	0	0	580	73	98.3

(a) The egg yolk used for purification was 100 g
(b) Total IgY (mg) / Total protein (mg) x 100
Hatta et al. (1990) [11], reproduced by permission of Japan Society for Bioscience, Biotechnology, and Agrochemistry, Tokyo.

C. SOME PROPERTIES OF PURIFIED IgY

The IgY isolated by the method described above was homogeneous in the test by gel chromatography. Figure 7 is the electrophoretic pattern of the IgY purified by the method of the SDS-PAGE system. The IgY was dissociated into a heavy chain and a light chain (64 kDa and 28 kDa) on SDS-PAGE under reduced condition. This electrophoretic pattern was quite similar to that shown by IgG isolated from chicken blood serum. Treating egg yolk with λ-carrageenan at the first step of purification was also found to serve for removal of several kinds of other yolk proteins. Chromatography on DEAE-Sephacel of the supernatant obtained after treating egg yolk with λ-carrageenan resulted in further purification of IgY. Nevertheless, the eluate from the column usually contains some other proteins which are presumed to be α– and β–livetins. These contaminating proteins could be removed by repeated salting-out with sodium sulfate under the conditions described above.

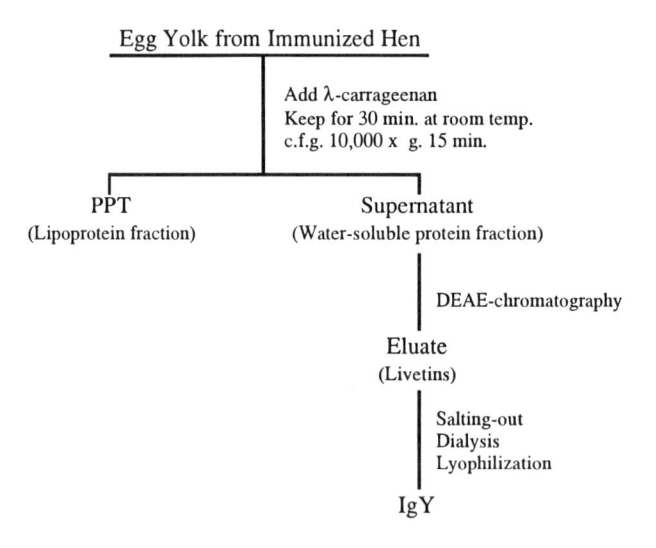

```
Egg Yolk from Immunized Hen
              |
        Add λ-carrageenan
        Keep for 30 min. at room temp.
        c.f.g. 10,000 x g. 15 min.
         /              \
       PPT            Supernatant
(Lipoprotein fraction)  (Water-soluble protein fraction)
                            |
                     DEAE-chromatography
                          Eluate
                         (Livetins)
                            |
                       Salting-out
                       Dialysis
                       Lyophilization
                          IgY
```

Figure 6. Purification of IgY through treatment with λ-carrageenan, DEAE-chromatography, and salting-out with sodium sulphate.

IV. SOME DIFFERENCES IN PROPERTIES BETWEEN IgG AND IgY

A. STRUCTURE AND IMMUNOLOGICAL FUNCTIONS

IgG in egg yolk, referred to as IgY, is somewhat different from the IgG of mammals in molecular weight, isoelectric point, binding behavior with complements, etc. The difference in properties between IgY and mammalian IgG has been reported as follows:

(1) The molecular weight of IgY is 1.8×10^4 Daltons while that of IgG is 1.5×10^4,
(2) The isoelectric point of IgY is lower than that of IgG by one pH unit [13],
(3) IgY does not associate with mammalian complements [14],
(4) The binding activity of IgY with the Fc receptor on the cell surface is much less than that by IgG [14],
(5) Unlike IgG, IgY never binds the *Staphylococcus* protein A [15],
(6) Unlike IgG, IgY does not bind rheumatoid factor in blood which is a marker of inflammatory response [16].

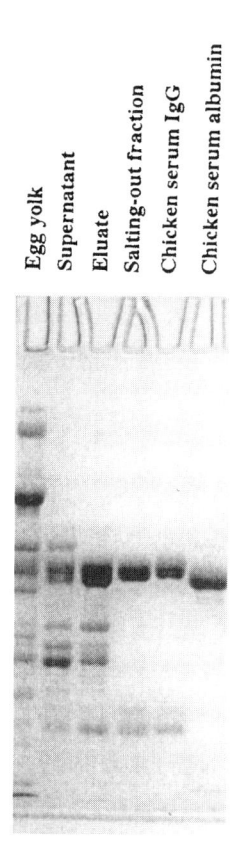

Figure 7. Changes in electrophoretic patterns of egg yolk proteins along with purification of IgY. (Refer to Table 2). From Hatta et al. (1990) [11] by permission of Japan Society for Bioscience, Biotechology, and Agrochemistry, Tokyo.

Recently, hen's gene to express IgG was cloned, and the constant regions of the heavy chain of IgY expressed was found to consist of four domains unlike that of mammalian IgG which consists of three domains (Figure 8) [17-19]. The molecular structure of IgY, therefore, is similar to mammalian IgM or IgE, which consist of four domains.

Very recently, Shimizu and his co-workers reported that the content of β–sheet structure in the constant domains of IgY was presumed to be lower than that of rabbit IgG, and the flexibility of the hinge region of IgY was greater than that of rabbit IgG [19]. Ohta and his co-workers [20] have made clear the whole structure of the various sugar chains of IgY, indicating that 27.1% of asparagine-linked carbohydrate chains of IgY have glucose as the nonreducing end residue of the glycolic chains, as described in Chapter 7.

B. HEAT AND pH STABILITY

The heat and pH stability of IgY and rabbit IgG specific to human rotavirus were compared by measuring the antibody activity by ELISA [21]. IgY was found to be significantly more sensitive than rabbit IgG at temperatures higher than 70°C (Figure 9). The study also showed that the temperature corresponding to the maximum of the denaturation endotherm (Tmax) of IgY was 73.9 °C while that of rabbit IgG was 77.0 °C analyzed with differential scanning calorimetry. The stability of IgY activity under various acidic conditions (pH 2 and 3) was

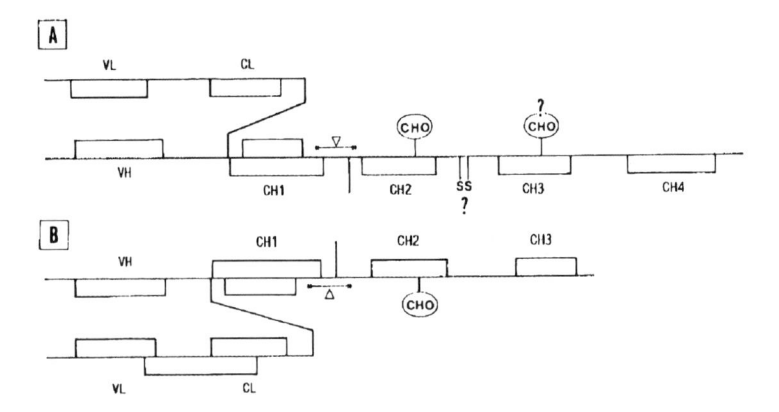

Figure 8. Difference in structure between IgY (A) and mammalian IgG (B). Reproduced from Shimizu et al. (1992) [19] by permission of Japan Society for Bioscience, Biotechology, and Agrochemistry, Tokyo.

more sensitive than that of rabbit IgG (Figure 10). Similar results have been reported by Otani and his co-workers [22] in the experiment compared using anti-α_{s1} casein IgY and rabbit IgG and Shimizu and his co-workers [23] using IgY and rabbit IgG specific to mouse IgG. This discrepancy in heat and acidic pH stability between IgY and IgG may be attributed to the difference in their protein structures.

C. STABILITY TO SEVERAL PROTEOLYTIC ENZYMES

Ohtani and his colleagues [22] reported that IgY specific to α_{s1}-casein was more susceptible to pepsin than rabbit IgG, but both antibodies were relatively stable to trypsin or chymotrypsin. Shimizu and his colleagues [24] also reported that the activity of IgY specific to *E. coli* examined by competitive ELISA was quite stable on incubation with trypsin or chymotrypsin but it was fairly sensitive to pepsin especially at pH values lower than 4.5. We also examined the stability of the IgY specific to human rotavirus against proteolytic enzymes by determination of the neutralization titer [25]. The activity estimated by neutralization titer of IgY was almost lost on incubation with pepsin at pH 2.0. But at pH 4.0, 91% of the activity remained after 1 h incubation with the enzyme. Even after 4 h incubation under this condition, 63% of the remaining activity was observed when examined by neutralization titer (Table 3). SDS-PAGE profiles of IgY after incubation with pepsin revealed that IgY at pH 2.0 was hydrolyzed into small peptides, and no bands corresponding to IgY were detected in the test by a linear-gradient polyacrylamide gel electrophoresis (Figure 11). On the contrary, in the incubation with pepsin at pH 4.0, heavy(H)- and light(L)-chain were clearly observed after 4 h, although some new bands appeared between H- and L-chains.

The behavior of IgY with trypsin and chymotrypsin was also examined. Changes in the neutralization titer of IgY were almost the same for the incubation with trypsin and with chymotrypsin. After 8 h incubation, the activity of IgY in neutralization titer remained 39% and 41% for the mixtures with trypsin and chymotrypsin, respectively (Table 4). SDS-PAGE profiles of IgY incubated with trypsin or chymotrypsin are shown in Figure 12. On incubation with trypsin, the IgY H-chain disappeared, and several bands between H- and L-chain appeared on SDS-PAGE. In the case with chymotrypsin, both H- and L-chains of IgY remained unchanged, although a small band below H-chain was observed.

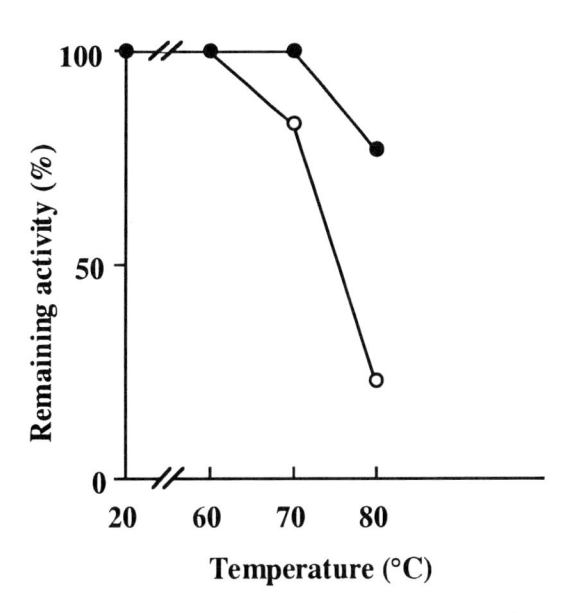

Figure 9. Comparison of heat stabilities of anti-HRV (Wa) IgY (○) and rabbit IgG (●). Remaining activity after 10 min heating at certain temperatures as measured by ELISA and expressed as a percent of the initial activity. Reproduced from Hatta et al. (1993) [21] by permission of Japan Society for Bioscience, Biotechology, and Agrochemistry, Tokyo.

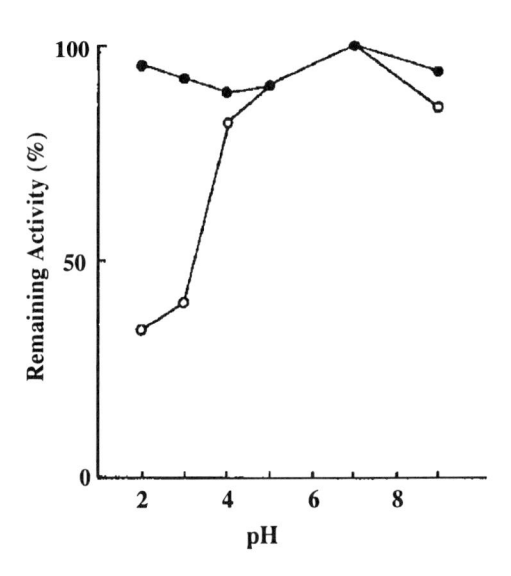

Figure 10. Comparison of pH stabilities of anti-HRV (Wa) IgY (○) and rabbit IgG (●). Remaining activity after heating (37°C, 1 h) at certain pH values as measured by ELISA and expressed as a percent of the initial activity. Reproduced from Hatta et al. (1993) [21] by permission of Japan Society for Bioscience, Biotechology, and Agrochemistry, Tokyo.

Table 3

Change in Nutralization Titer of Anti-HRV(MO) IgY during Heating with Pepsin [a]

Incubation period (h)	pH 2.0 Titer	Recovery (%)	pH 4.0 Titer	Recovery (%)
0	3,200	100	3,200	100
1	<10	0	2,900	91
2	<10	0	2,300	72
4	<10	0	2,000	63

[a] The ratio of enzyme: IgY by weight was 1 : 200, and the mixture was heated at 37°C. Reproduced from Hatta et al. (1993) [25] by permission of Japan Society for Bioscience, Biotechology, and Agrochemistry, Tokyo.

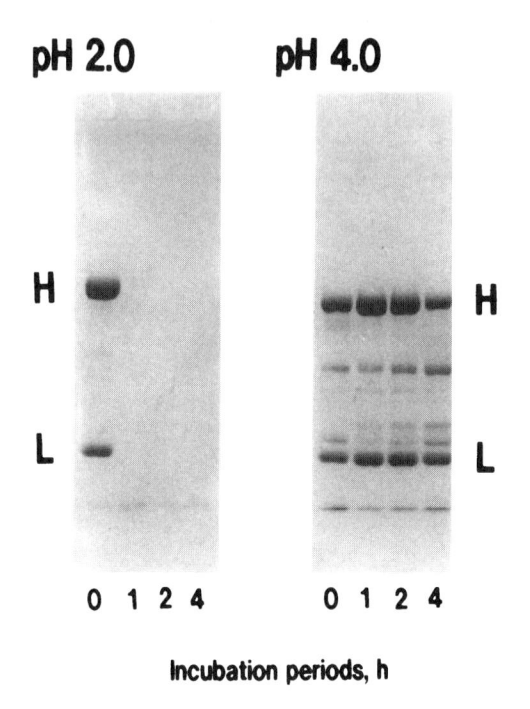

Figure 11. SDS-PAGE profiles of anti-HRV (MO) IgY incubated with pepsin at pH 2.0 and 4.0. H and L indicate heavy and light chains of IgY, respectively. Reproduced from Hatta et al. (1993) [25] by permission of Japan Society for Bioscience, Biotechology, and Agrochemistry, Tokyo.

Table 4

Change in Neutralization Titer of Anti-HRV(MO) IgY
during Heating with Trypsin or Chymotrypsin [a]

Incubation period (h)	Trypsin Titer	Recovery (%)	Chymotrypsin Titer	Recovery (%)
0	5,600	100	5,600	100
1	4,500	80	4,000	71
4	3,700	66	3,000	54
8	2,200	39	2,300	41

[a]The ratio of enzyme: IgY by weight was 1 : 20, and the mixture was heated at 37°C. Reproduced from Hatta et al. (1993) [25] by permission of Japan Society for Bioscience, Biotechology, and Agrochemistry, Tokyo.

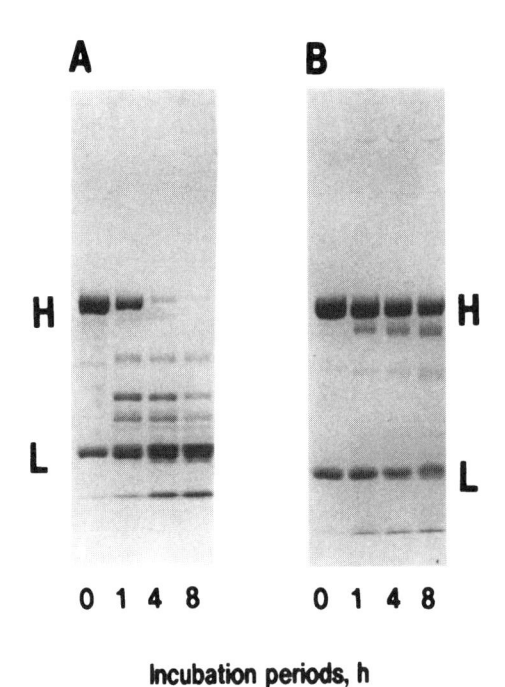

Figure 12. SDS-PAGE profiles of anti-HRV (MO) IgY incubated with trypsin (A) and chymotrypsin (B). H and L indicates heavy and light chains of IgY, respectively. Reproduced from Hatta et al. (1993) [25] by permission of Japan Society for Bioscience, Biotechology, and Agrochemistry, Tokyo.

V. APPLICATION OF IgY AS AN IMMUNOLOGICAL TOOL

Antigen-specific IgG isolated from sera of superimmunized animals, such as rabbits, cows, and goats has been widely applied as an immunological tool in the field of diagnosis as well as pure research. The antigen-specific IgY is useful in its binding specificity as well as mammalian IgG is specific to given antigens. They both serve to detect antigens for which specificity will never be achieved by any other method.

A. DIAGNOSIS

Altschuh and his co-workers [26] reported that IgY against human antibody (IgG and IgM) was applicable to determining their concentration in biological fluid by the method of rocket-immunoelectrophoresis. In the case of using rabbit IgG in this method, chemical modification, such as carbamylation, of the IgG is generally needed to change its isoelectric point from that of the human antibody. However, in application of IgY by this method, the carbamylation was not necessary since its isoelectric point is different from that of the human antibody. Fertel and his co-workers [27] demonstrated the application of IgY in determining prostaglandin in serum using radioimmunoassay in which they used prostaglandin conjugated with hemocyanin (hapten) as an antigen for immunization of hens. Gardner and Kaya [14] prepared IgY specific to rotavirus, adenovirus, and influenza virus, demonstrated immunological detection of these viruses using the IgY as the first antibody and fluorescein isothiocyanate (FITC)-conjugated rabbit IgG specific to IgY as the second antibody. They suggested that the preparation of IgY specific to these viruses was achieved with much more convenience over the conventional rabbit IgG, because of no necessity of purification of the virus as antigen. Since these viruses can be cultivated using fertilized eggs, the contaminants in the virus culture that are components of egg must not show any immunogenicity to hens as far as hens are immunized with the virus culture as an antigen. They also suggested that IgY is a suitable antibody for detecting pathogens in stool samples, because it does not bind protein A derived from *Staphylococcus aureus* usually found in stool. Owing to this property of IgY, the fault of the detection method of pathogens in stool can be avoided. Many researchers have also demonstrated the application of IgY for determination of various important, but very minor biological substances, such as plasma kallikrein [28], 1, 25-dihydroxyvitamin D [29], hematoside (NeuGc) [30], human transferrin [31], ochratoxin A [32], human dimeric IgA [33], and high-molecular weight mucin-like glycoprotein-A (HMGP-A) [34].

Another advantage in the use of IgY as an immunological tool over using rabbit IgG is the sensitivity of hens to antigens originated from mammalians. A number of proteins exist whose amino acid sequence are well preserved among mammals, and many of these proteins have no or little antigenicity toward mammals. Therefore, for preparing antibodies against those proteinous antigens, the hen is highly promising as an alternative animal, because of the immunological distance of hen from mammalians. In fact, Carroll and Stollar [35] succeeded in preparing IgY against RNA polymerase II which has not generated its specific antibody in mammals. Many researchers have succeeded in producing IgY specific to less immunogenic antigen against mammals, such as heat-shock protein (Hsp70) [36], peptidylglycine α-amidating enzyme [37], parathyroid hormone related protein [38], proliferating cell nuclear antigen of calf thymus [39], von Willebrand factor [40], and platelet glycoprotein IIb-IIIa [40]. Our group also succeeded in the preparation of IgY against rat glutathion peroxidase [41], human insulin [42], and mouse erythropoietin receptor [43] which have significantly less antigenicity among mammals.

B. IgY AS A LIGAND OF AN IMMUNOADSORBENT

Immuno-affinity chromatography has been applied as a useful method for purification of proteins (antigens). Rabbit IgG has been conventionally used as a ligand to be immobilized to the adsorbent, such as cellulose or agarose. However, several disadvantages of this affinity chromatography have been pointed out when rabbit IgG was used as a ligand, because acidic pH values less than pH 2 are necessary for dissociation of the protein immunologically bound to the rabbit IgG on the immunoadsorbent. Therefore, the dissociated protein is often denatured depending on its nature. Moreover, production of rabbit IgG in large amounts is generally expensive. The immuno-affinity chromatography using rabbit IgG as an ligand has thus been applied for isolation of only certain specific proteins.

Our research group demonstrated recently that IgY is an effective alternative antibody as a ligand for an immunoadsorbent. In the experiment, IgY and rabbit IgG specific to mouse IgG were immobilized on Sepharose 4B, respectively, in order to compare its dissociation efficiency for the purification of mouse IgG. Mouse serum was applied on immunoadsorbents, and the adsorbent was eluted with the buffer solution of pH 4.0 and 2.0, stepwise. The mouse IgG dissociated at pH 4.0 was only half of that applied, and the remaining IgG was eluted with pH 2.0 buffer solution in an immunoadsorbent using rabbit IgG as a ligand. On the other hand, 97% of the mouse IgG was dissociated even at pH 4.0 on the immunoadsorbent using IgY as a ligand.

VI. PASSIVE IMMUNIZATION BY USE OF IgY

Another important application of IgY is for passive immunization therapy in which the specific binding ability to the antigens (pathogens, venoms, etc.) serves to neutralize the biological activities of those antigens. Passive immunization seems to be one of the most valuable applications of antibodies in which pathogen-specific IgG is administered to individuals to result in prevention of infectious diseases. Passive immunization differs from active immunization (vaccination) in that the former employs an antibody obtained from other animals. The administration of this antibody specific to certain antigens (bacteria, virus, toxin, etc.) to individuals orally or systemically works to neutralize infectious activity or toxicity of the antigens. For practical application of passive immunization, an effective method of preparation of the antibody will be necessary, because large amounts of antibody may be required to administer the antibody for the passive immunization. The antigen-specific IgY can now be prepared on an industrial scale from eggs laid by hens immunized with selected antigens. Passive immunization using IgY will be widely practiced in the near future.

A. SYSTEMIC ADMINISTRATION OF IgY
1. Prevention of Newcastle disease virus

Newcastle disease has now been controlled well by immunizing hens with an attenuated Newcastle disease virus (vaccination). In the vaccination (active immunization), it takes about a week before individuals show sufficient antibody activity in their sera for prevention of virus infection. Therefore, a week after vaccination is a risk period during which active antibodies are generated. Since IgY specific to Newcastle disease virus is produced from egg yolk, and the use of the IgY to hens might not have immunogenicity, there is an idea that a risk period could be covered by intramuscular administration of the IgY to hens at the same time hens are vaccinated. Stedman and his colleague [44] demonstrated the effectiveness of IgY isolated from egg yolk in conferring passive protection against challenge by Newcastle disease virus.

2. Neutralization of venom

It has been estimated that 1.7 million people are bitten by venomous snakes, scorpions, jelly fish, and spiders, and 40,000 to 50,000 die annually. For treating the patient, antivenom horse serum (IgG) is administered. The antivenom horse serum administered to individuals neutralizes the fatal toxicity of the venoms and saves their lives. However, there have been several papers stating that horse IgG frequently causes human complement mediated side effects. Also, impurities existing in the horse serum may cause serum sickness or anaphylactic shock. Thalley and Carroll [45] demonstrated the usefulness of IgY specific to venoms for neutralizing its toxicity using animal models. They mentioned that IgY has no or little risk of causing complement mediated side effects because IgY has no activity to bind human complement. In the future, IgY specific to snake venom may be used instead of antivenom horse serum for therapy of patients affected by snake venom.

B. ORAL ADMINISTRATION OF IgY

The most attractive application of IgY is oral passive immunization therapy. At present, several researchers have reported that oral administration of IgY is quite effective in prevention of rotaviral diarrhea [25, 46-48], dental caries [9, 49], enterotoxigenic *E. coli* infection [50] and infectious disease of fish [51, 52]. In this section, we describe our results regarding oral passive immunization using IgY.

1. Prevention of rotaviral diarrhea

Human rotavirus (HRV) was found in 1973 by Bishop and his colleagues [53] and is known as a major pathogen of infectious gastroenteritis in infants and young children. HRV infection is characteristically localized in the epithelial cells of the intestinal tract and causes severe diarrhea with vomiting. It has been reported that the deaths of infants attributed to HRV diarrhea disease, which occur mainly in developing countries, amount to more than two million annually [54]. Vaccination trials for HRV infection have remained unsuccessful because of the difficulty of introducing the active antibody to the intestinal tract of the infants whose immunity generally has not developed yet [55]. Prevention of HRV infectious disease by oral administration of the active antibody (passive immunization therapy) may be a promising application of anti-HRV antibody. This is because the infection with HRV is limited to epithelial cells of the intestinal tract. Oral passive immunization has thus been investigated as an alternative to vaccination for prevention of HRV infectious disease. Several researchers demonstrated that oral administration of antirotavirus IgG or IgY was effective in controlling rotaviral diarrhea using several animal models [46-48, 56-58].

Our experiment showed that immunization of hens with HRV results in bringing high levels of titer in yolk of the egg laid by hens. The antibody titer of high level is usually observed in any egg laid over a year after immunization [21]. The anti-HRV IgY was purified and the HRV challenge test was undertaken using suckling mice according to the method by Ebina and his colleagues [48]. Suckling mice (11-16 mice/group) were inoculated with HRV MO strain (3.5×10^7 FCFU/mouse) at 1, 3, 9, and 24 h after oral administration of anti-HRV(MO) IgY (225 mg IgY/mouse). No incidents of diarrhea were observed in all the mice that were administered IgY 1 h before HRV inoculation (Figure 13). However, the mice administered IgY at 3, 9, and 24 h before HRV inoculation suffered from diarrhea with incidence of 27.3, 41.7, and 93.3%, respectively. In the positive control group, the incidence of diarrhea was 83.3%. In our investigation, the preventive effect from HRV diarrhea in suckling mice was found to decrease as the time gap between IgY administration and HRV inoculation was lengthened. This result indicates that the effectiveness of IgY for prevention of HRV diarrhea depends greatly on the time when IgY is administered before infection with HRV [25].

Photographs of sectioned small intestines of suckling mice are shown in Figure 14. Morphologically, expansion and vacuolation of cells and destruction of villus tips were observed in the sectioned small intestines of suckling mice that were suffering severely from diarrhea by HRV infection (1.8x10^7 FCFU/mouse). However, the sections of those mice prevented from HRV diarrhea by oral administration of anti-HRV IgY (225 mg IgY/mouse) at 1 h before HRV inoculation (1.8x10^7 FCFU/mouse) were morphologically the same as those of noninoculated control mice. Also, anti-HRV(MO) IgY (225 mg/mouse) was orally administered to suckling mice (8-12 mice/group) at 1, 3, 10, and 24 h after inoculation of HRV MO strain (3.0x10^7 FCFU/mouse). All the mice of the positive control group suffered from diarrhea. On the other hand, the incidence of diarrhea of the mice which were given with the IgY at 1, 3, 10, and 24 h after HRV inoculation, was 37.5, 45.5, 41.2, and 70.0%, respectively (Figure 15). There have been no reports published on the therapeutic effect of IgY on HRV diarrhea induced in suckling mice. We demonstrated that oral administration of anti-HRV(MO) IgY within 24 h after HRV infection was very effective in decreasing the number of suckling mice suffering from HRV-induced diarrhea [25].

Ebina and his colleagues [59] have reported that oral administration of anti-HRV antibodies (skimmed colostrum) prevented several infants from getting rotavirally induced gastroenteritis.

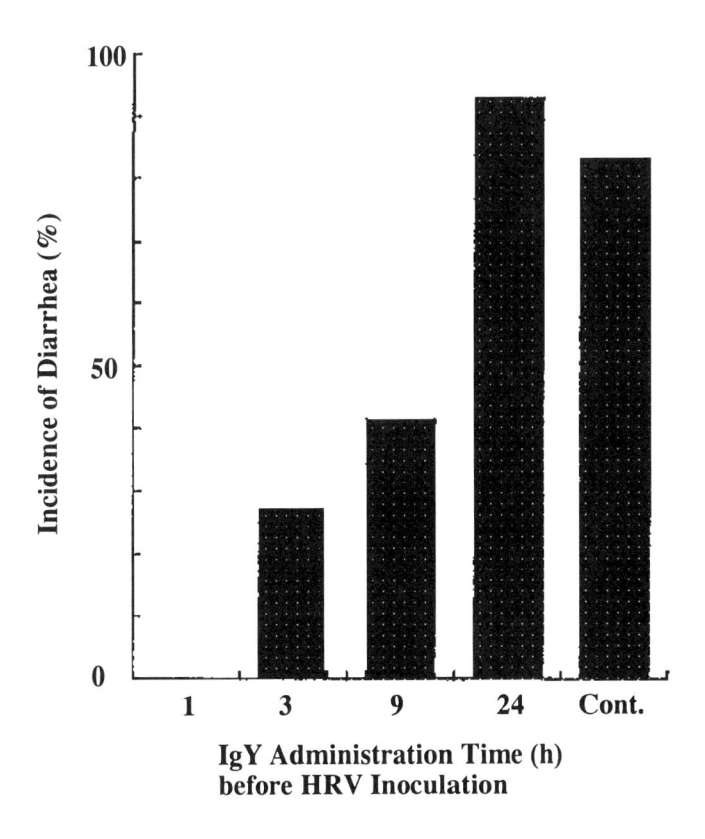

Figure 13. Passive immunization effect of anti-HRV IgY administered to suckling mice before inoculation of HRV. Reproduced from Hatta et al. (1993) [25] by permission of Japan Society for Bioscience, Biotechology, and Agrochemistry, Tokyo.

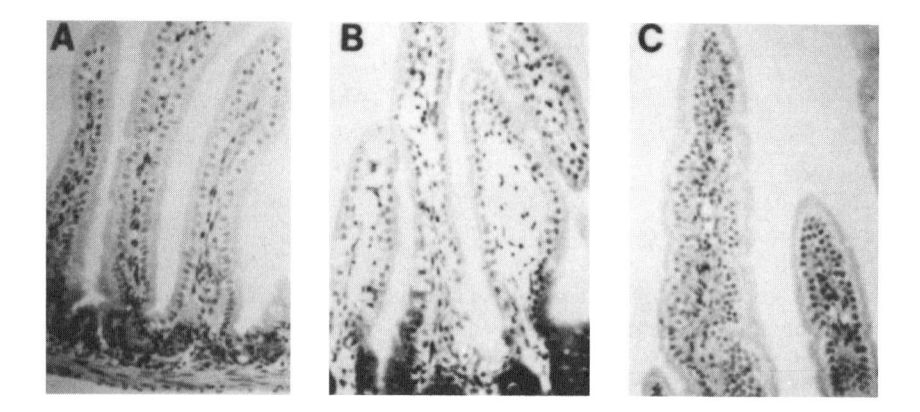

Figure 14. Histological observation of epithelial cells of the small intestine of suckling mice with or without infection with HRV: A) normal tissue; B) tissue infected by HRV; C) tissue of a mouse orally administered anti-HRV IgY. Reproduced from Hatta et al. (1993) [25] by permission of Japan Society for Bioscience, Biotechology, and Agrochemistry, Tokyo.

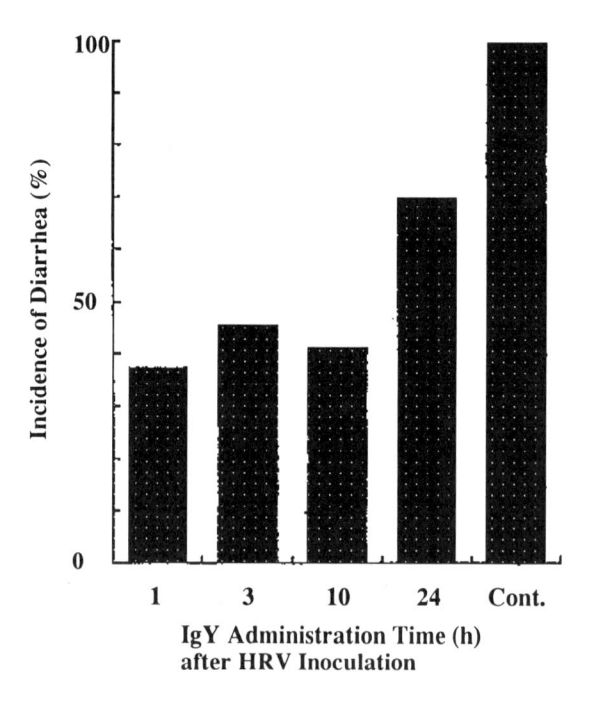

Figure 15. Passive immunization effects of anti-HRV IgY administered to suckling mice after inoculation of HRV. Reproduced from Hatta et al. (1993) [25] by permission of Japan Society for Bioscience, Biotechology, and Agrochemistry, Tokyo.

This fact suggests that HRV infection in infants can be prevent by oral administration of anti-HRV IgY, as long as the IgY is introduced in the small intestine, without any significant loss of activity by gastric enzymes. We revealed that anti-HRV IgY reaches the small intestine of suckling mice in a relatively short time after oral administration. Therefore, it is highly likely that oral administration of anti-HRV IgY at certain time intervals is practically effective for prevention of HRV-induced diarrhea in human infants.

2. Prevention of Fish Disease

The infection of eels with *Edwardsiella tarda* is known to cause the most severe mortality in eels among all their infectious diseases. The wide spread of the drug-resistant strain of *E. tarda* has recently been reported as giving rise to ineffectiveness of chemotherapy. The method of vaccination (active immunization) was also studied as an alternative control method for *E. tarda* infection. The method of intraperitoneal injection of formalin-killed *E. tarda* was found to be effective, inducing protective immunity for the eel. However, such methods may be actually impossible to practice in an eel cultivation farm.

We have demonstrated that Edwardsiellosis of Japanese eels can be prevented by oral administration of the anti-*E. tarda* IgY [51, 52]. For the immunization test, *E. tarda* (SH-89108) isolated from naturally diseased eels by Shizuoka Prefectural Experimental Fisheries Station in 1989 was used as the antigen. Changes in anti-*E. tarda* IgY levels of egg yolk were investigated by the ELISA method over 40 weeks after the initial immunization (Figure 16).

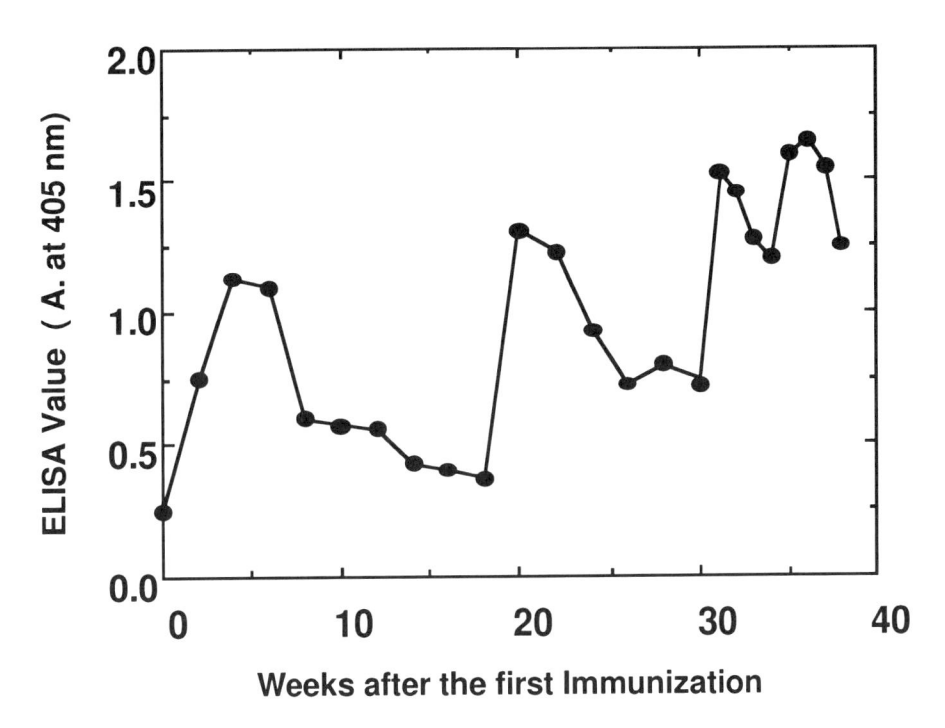

Figure 16. The course of IgY activity against *E. tarda*. Hens were immunized at 0, 1, 2, 3, 18, 30, and 34 weeks. Reproduced from Hatta et al. (1994) [52] by permission of CAB International, Oxon.

The immunization was done once a week. The level of anti-*E. tarda* IgY in the egg yolk greatly increased after the second injection and reached its highest level in the fifth week. Thereafter, the IgY level gradually decreased. But, when an additional immunization was performed, the active IgY again reached its maximum level. Thus, a shorter additional immunization intermittently was found to be effective in maintaining the high IgY level. It is noteworthy that no change in the egg-laying rate of hens was observed during the immunization periods.

The egg yolk from *E. tarda*–superimmunized hens (5-8 weeks after the initial immunization) was separated into lipoprotein and water-soluble protein fractions by the carrageenan method. The water-soluble protein fraction was filtered through a filter paper and lyophilized. This lyophilized preparation was used as anti-*E. tarda* IgY in the experiments. Japanese eels (ca. 200 g body weight) were anesthetized in 1.5% urethane solution. A polyethylene tube of 3 cm with 1.0 mm diameter was inserted into the intestine through the anus and 0.1 ml hydrogen peroxide solution (30%) was infused to cause damage in the intestinal mucosa. After 18 to 24 h, 10^5-10^6 CFU/eel of viable *E. tarda* was mixed with a kneaded feed (2.4 g) and administered into the eel stomach by cannula under anesthesia. In the experiment for the effect of anti-*E. tarda* IgY, the IgY (400 mg/eel) was mixed into the feed. These eels were kept in a tank containing 30 l water at 25°C for 40 days and the mortalities were assessed. The anatomy and histopathology of the dead or moribund eels were performed for accumulations of abscess formation due to Edwardsiellosis. Administration of *E. tarda* (10^5-10^6 cells/eel) brought about severe mortalities to the eels within 10 days (Figure 17). All the dead or surviving eels showed abscess formation in the liver or kidney indicating that they had Edwardsiellosis. On the other hand, the eels administrated anti-*E. tarda* IgY (400 mg/eel) together with *E. tarda* (10^5-10^6 cells/eel) survived over the experimental period of 40 days, without showing any symptom of Edwardsiellosis. The eels previously treated with hydrogen peroxide infusion (surgical control) also survived throughout the experimental period.

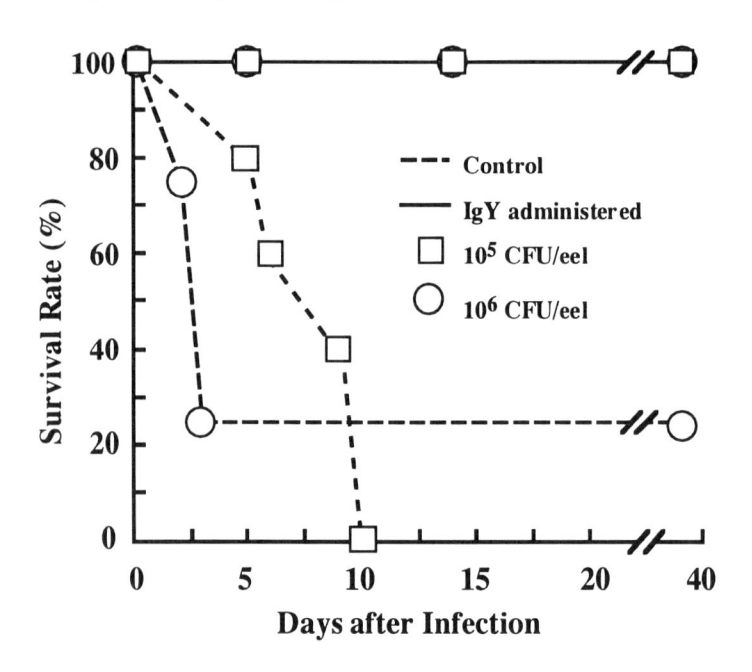

Figure 17. Preventive effect of IgY against *E. tarda* infection in Japanese eels. Reproduced from Hatta et al. (1994) [52] by permission of CAB International, Oxon.

Many pathogens of fish have been reported to spread by infection through intestinal mucosa. The oral supply of specific IgY against fish pathogens in feed will be an alternative to the method of using antibiotics or chemotherapeutics for prevention of fish diseases. Moreover, the oral supply of active IgY would be a novel prevention method against viral infectious diseases of fish, because no drugs have so far been developed as effective substances against those viral fish diseases.

3. Prevention of Dental Caries

Streptococcus mutans serotype *c* is thought to be the principal causative bacterium of dental caries in humans [60-62]. Several studies have been undertaken regarding preventive measures for protecting the host from such infectious disease. Many reports suggested the possibility of preventing dental caries by vaccination (active immunization) using *S. mutans* whole cells or one of its characteristic cariogenic factors as an antigen. Also, many researchers reported the effectiveness of passive immunization against caries in which specific antibodies against *S. mutans* were administered orally. Recently, passive immunization has gained much attention because active immunization may cause positive side effects caused by *S. mutans* vaccine antigens.

We prepared IgY specific to *S. mutans* serotype c and demonstrated its effect on prevention of dental caries in the passive immunization test using rats [49]. The *S. mutans* MT8148 (serotype *c*) cultivated in the medium containing sucrose was used as an antigen. Fifty hens (150 days old) were immunized with 2 ml of the antigen (containing 1×10^9 CFU killed *S. mutans*) by intramuscular injection into both legs. The immunization was repeated once a

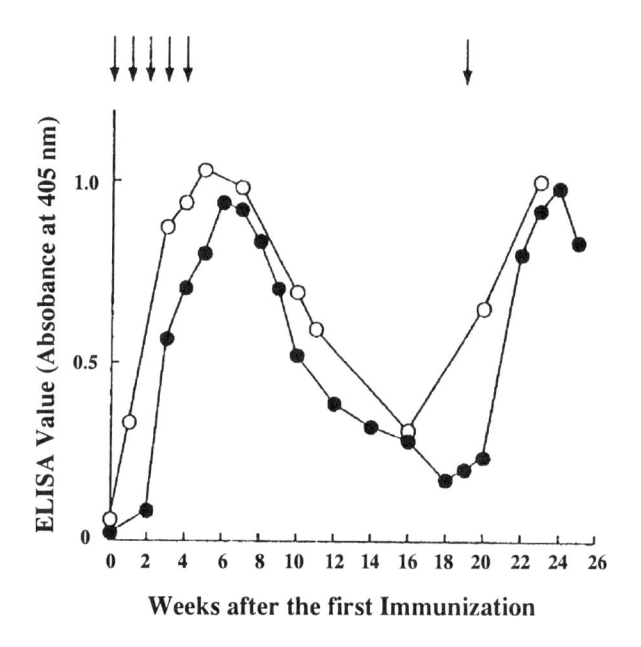

Figure 18. Egg yolk and serum antibody values against *S. mutans*. Closed and open circles indicate antibody levels in egg yolk and serum, respectively, of hens immunized with *S. mutans* MT8148. Arrows indicate the immunization schedule. Intramuscular immunization was performed once a week for four weeks after the initial immunization. Boosting was also done at 19 weeks. Reproduced from Otake et al. (1991) [49] by permission of International-American Association for Dental Research, Alexandaria, VA.

week for four weeks, and an additional immunization was conducted when the antibody titer decreased. Changes of antibody levels in both egg yolk and hen's serum against *S. mutans* antigen were investigated by ELISA for 25 weeks after the initial immunization (Figure 18). The antibody levels in the egg yolk increased in one or two weeks after the antibody appeared in the serum. Although the levels decreased gradually, the antibody levels in both serum and egg yolk were restored by the booster immunization at the 18th week. There was no decrease in the egg-laying rate during the immunization periods.

For the preparation of immune yolk powder, egg yolk separated from eggs of immunized hens (laid between 4 and 10 weeks after initial immunization) was homogenized and spray-dried. Control yolk powder was also prepared from eggs laid by nonimmunized hens. The experimental rat caries model was used in order to test the anticariogenic effect of immune yolk powder containing IgY specific to *S. mutans*. Specific pathogen-free rats infected with *S. mutans* MT8148 and fed with a cariogenic diet containing more than 2% immune yolk powder showed significantly less caries scores than those infected with the same strain and fed with a diet containing control yolk powder (Table 5). It has been demonstrated that adsorption of *S. mutans* on saliva–coated hydroxyapatite is remarkably inhibited by pretreatment of *S. mutans* with the IgY purified from the immune yolk powder. Therefore, we suggested that a part of the rat caries reduction shown by the IgY was due to the inhibition of *S. mutans* adhesion onto the teeth surface.

Recently, researchers have reported the possibility of preventing caries by passive immunization using mouse monoclonal IgG against *S. mutans* cell-surface antigen [63, 64] and immunized bovine whey IgG against *S. mutans* whole cells [65]. Our study provides additional evidence that anti-*S. mutans* IgY prepared from eggs can be used for passive immunization to prevent the development of dental caries.

Table 5

Mean Caries Scores in Mandibular Molars of Specific Pathogen-free Rats Infected with *S. mutans* MT 8148 (c) and Fed with Diet M2000 for 58 Days

Groups (Ratio of Immune/Control Yolk Powder in Diet)	Mean Caries Score ($\pm$ SE)			
	Sulcal	Buccal	Approximal	Total
A(100/0)	40.1$\pm$2.5*	2.1$\pm$0.3*	0.5$\pm$0.3*	42.8$\pm$2.8*
B(33/67)	64.1$\pm$3.7*	5.6$\pm$0.4**	1.4$\pm$0.6*	71.1$\pm$4.1*
C(10/99)	71.1$\pm$5.0*	7.3$\pm$1.2	3.0$\pm$0.8	81.4$\pm$6.5*
D(0/100)	90.4$\pm$1.1	9.6$\pm$1.1	4.9$\pm$0.8	104.9$\pm$1.9

56% sucrose in diet 2000 was replaced by 36% sucrose and 20% egg yolk powder containing different ratios of immune and control yolk powders. Different ratios of egg yolk powder prepared from immunized and normal hens were added to diet 2000. For example, diet M2000 used in group B contained a ratio of 33 to 67 of immune and control yolk powder, respectively, in the 20% egg yolk powder. Each group contained seven rats. The average weight of the rats was 234.1 $\pm$3.2 g, and the weights of rats in the four different groups were very similar. Statistical analyses (t tests) were carried out between group D and the other groups.
*Significance of difference, $p<0.01$.
**Significance of difference, $p<0.05$.
Reproduced from Otake et al. (1991) [49] by permission of International-American Association for Dental Research, Alexandaria, VA.

REFERENCES

1. **Leslie, G. A. and Clem, L. W.**, Phylogeny of immunoglobulin structure and function. III. Immunoglobulins of the chicken, *J. Exp. Med.*, **130**, 1337, 1969.
2. **Leslie, G. A. and Martin, L. N.**, Studies on the secretory immunologic system of fowl. III. Serum and secretory IgA of the chicken, *J. Immunol.*, **110**, 1, 1973.
3. **Rose, M. E., Orlans, E., Buttress, N.**, Immunoglobulin classes in the hen's egg: Their segregation in yolk and white, *Eur. J. Immunol.*, **4**, 521, 1974.
4. **Jensenius, J. C., Andersen, I., Hau, J., Crone, M., and Koch, C.**, Eggs: Conveniently packaged antibodies. Methods for purification of yolk IgG, *J. Immunol. Methods*, **46**, 63, 1981.
5. **Gottstein, B. and Hemmeler, E.**, Egg yolk immunoglobulin Y as an alternative antibody in the serology of echinococcosis, *Z. Parasitenkd*, **71**, 273, 1985.
6. **McBee, L. E. and Cotterill, O. J.**, Ion-exchange chromatography and electrophoresis of egg yolk proteins, *J. Food Sci.*, **44**, 656, 1979.
7. **Bade, H. and Stegemann, H.**, Rapid method of extraction of antibodies from hen egg yolk, *J. Immunol. Methods*, **72**, 421, 1984.
8. **Polson, A., Coetzer, T., Kruger, J., Malitzahn, E., and Merwe, K. J.**, Improvements in the isolation of IgY from the yolks of eggs laid by immunized hens, *Immunol. Invest.*, **14**, 323, 1985.
9. **Hamada, S., Horikoshi, T., Minami, T., Kawabata, S., Hiraoka, J., Fujiwara, T., and Ooshima, T.**, Oral passive immunization against dental caries in rats by use of hen egg yolk antibodies specific for cell-associated glucosyltransferase of *Streptococcus mutans*, *Infect. Immun.*, **59**, 4161, 1991.
10. **Hatta, H., Sim, J. S., and Nakai, S.**, Separation of phospholipids from egg yolk and recovery of water-soluble proteins, *J. Food Sci.*, **53**, 425, 1988.
11. **Hatta, H., Kim, M., and Yamamoto, T.**, A novel isolation method for hen egg yolk antibody, "IgY", *Agric. Biol. Chem.*, **54**, 2531, 1990.
12. **Hassl, A., Aspoch, H., and Flamm, H.**, Comparative studies on the purity and specificity of yolk immunoglobulin Y isolated from eggs laid by hens immunized with *Toxoplasma gondii* antigen, *Zbl. Bakt. Hyg.*, **A267**, 247, 1987.
13. **Polson, A., Wechmar, M. B. V., and Fazakerley, G.**, Antibodies to proteins from yolk of immunized hens, *Immunol. Commun.*, **9**, 495, 1980.
14. **Gardner, P. S. and Kaye, S.**, Egg globulins in rapid virus diagnosis, *J. Virol. Methods*, **4**, 257, 1982.
15. **Kronvall, G., Seal, U. S., Svensson, S., and Williams, R. C. Jr.**, Phylogenetic aspects of staphylococcal protein A-reactive serum globulins in birds and mammals, *Acta. Path. Microbiol. Scand. Section B*, **82**, 12, 1974.
16. **Larsson, A. and Sjoquist, J.**, Chicken antibodies: A tool to avoid false positive results by rheumatoid factor in latex fixation tests, *J. Immunol. Methods*, **108**, 205, 1988.
17. **Parvari, R., Avivi, A., Lentner, F., Ziv, E., Tel-Or, S., Burstein, Y., and Schechter, I.**, Chicken immunoglobulin γ-heavy chains: Limited V_H gene repertoire, combinatorial diversification by D gene segments and evolution of the heavy chain locus, *EMBO J.*, **7**, 739, 1988.
18. **Reynaud, C. A., Dahan, A., Anquez, V., and Weill, J. C.**, Somatic hyperconversion diversifies the single VH gene of the chicken with a high incidence in the D region, *Cell*, **59**, 171, 1989.
19. **Shimizu, M., Nagashima, H., Sano, K., Hashimoto, K., Ozeki, M., Tsuda, K., and Hatta, H.**, Molecular stability of chicken and rabbit immunoglobulin G, *Biosci. Biotech..*

Biochem., **56,** 270, 1992.
20. **Ohta, M., Hamako, J., Yamamoto, S., Hatta, H., Kim, M., Yamamoto, T., Oka, S., Mizuochi, T., and Matuura, F.,** Structures of asparagine-linked oligosaccharides from hen egg-yolk antibody (IgY). Occurence of unusual glucosylated oligomannose type oligosaccharides in a mature glycoprotein, *Glycoconjugate J.,* **8,** 400, 1991.
21. **Hatta, H., Tsuda, K., Akachi, S., Kim, M., and Yamamoto, T.,** Productivity and some properties of egg yolk antibody (IgY) against human rotavirus compared with rabbit IgG, *Biosci. Biotech. Biochem.,* **57,** 450, 1993.
22. **Otani, H., Matsumoto, K., Saeki, A., and Hosono, A.,** Comparative studies on properties of hen egg yolk IgY and rabbit serum IgG antibodies, *Lebensm. Wiss. u. Technol,* **24,** 152, 1991.
23. **Shimizu, M., Nagashima, H., and Hashimoto, K.,** Comparative studies on molecular stability of immunoglobulin G from different species, *Comp. Biochem. Physiol.,* **106B,** 255, 1993.
24. **Shimizu, M., Fitzsimmons, R. C., and Nakai, S.,** Anti-*E. coli* immunoglobulin Y isolated from egg yolk of immunized chickens as a potential food ingredient, *J. Food Sci.,* **53,** 1360, 1988.
25. **Hatta, H., Tsuda, K., Akachi, S., Kim, M., Yamamoto, T., and Ebina, T.,** Oral passive immunization effect of antihuman rotavirus IgY and its behavior against proteolytic enzymes, *Biosci. Biotech. Biochem.,* **57,** 1077, 1993.
26. **Altschuh, D., Hennache, G., and Regenmortel, M. H. V.,** Determination of IgG and IgM levels in serum by rocket immunoelectrophoresis using yolk antibodies from immunized chickens, *J. Immunol. Methods,* **69,** 1, 1984.
27. **Fertel, R., Yetiv, J. Z., Coleman, M. A., Schwarz, R. D., Greenwald, J. E., and Bianchine, J. R.,** Formation of antibodies to prostaglandins in the yolk of chicken eggs, *Biochem. Biophys. Res. Commun.,* **102,** 1028, 1981.
28. **Burger, D., Ramus, M. A., and Schapira, M.,** Antibodies to human plasma kallikrein from egg yolks of an immunized hen: Preparation and characterization, *Thromb. Res.,* **40,** 283, 1985.
29. **Bauwens, R. M., Devos, M. P., Kint, J. A., and Leenheer, A. P.,** Chicken egg yolk and rabbit serum compared as sources of antibody for radioimmunoassay of 1,25-dihydroxyvitamin D in serum or plasma, *Clin. Chem.,* **34,** 2153, 1988.
30. **Hirabayashi, Y., Suzuki, T., Suzuki, Y., Taki, T., Matumoto, M., Higashi, H., and Kato, S.,** A new method for purification of antiglycosphingolipid antibody. Avian antihematoside (NeuGc) antibody, *J. Biochem.,* **94,** 327, 1983.
31. **Ntakarutimana, V., Demedts, P., Sande, M. V., and Scharpe, S.,** A simple and economical strategy for downstream processing of specific antibodies to human transferrin from egg yolk, *J. Immunol. Methods,* **153,** 133, 1992.
32. **Clarke, J. R., Marquardt, R. R., Frohlich, A. A., Oosterveld, A., and Madrid, F. J.,** Isolation, characterisation, and application of hen egg yolk polyclonal antibodies, in *Egg Uses and Processing Technologies-New Developments,* Sim, J. S., and Nakai, S., Eds., CAB International, Oxon, 1994, p 207.
33. **Polson, A., Maass, R., and Van Der Merwe, K. J.,** Eliciting antibodies in chickens to human dimeric IgA removal of factors from human colostrum depressing anti IgA antibody production, *Immunol. Invest.,* **18,** 853, 1989.
34. **Shimizu, M., Watanabe, A., and Tanaka, A.,** Detection of high-molecular weight mucin-like glycoprotein-A (HMGP-A) of human milk by chicken egg yolk antibody, *Biosci. Biotech. Biochem.,* **59,** 138, 1995.
35. **Carroll, S. B. and Stollar, B. D.,** Antibodies to calf thymus RNA polymerase II from egg

yolks of immunized hens, *J. Biol. Chem.*, **258**, 24, 1983.
36. **Gutierrez, J. A. and Guerriero, V., Jr.**, Quantitation of Hsp 70 in tissues using a competive enzyme-linked immunosorbent assay, *J. Immunol. Methods,* **143**, 81, 1991.
37. **Sturmer, A. M., Driscoll, D. P., and Jackson-Matthews, D. E.**, A quantitative immunoassay using chicken antibodies for detection of native and recombinant a-amidating enzyme, *J. Immunol. Methods*, **146**, 105, 1992.
38. **Rosol, T. J., Steinmeyer, C. L., McCauley, L. K., Merryman, J. I., Werkmeister, J. R., Grone, A., Weckmann, M. T., Swayne, D. E., and Capen, C. C.**, Studies on chicken polyclonal anti-peptide antibodies specific for parathyroid hormone related protein (1-36), *Vet. Immunol. Immunopathol.*, **35**, 321, 1993.
39. **Gassmann, M., Thommes, P., Weiser, T., and Hubsgher, U.**, Efficient production of chicken egg yolk antibodies against a conserved mammalian protein, *FASEB J.*, **4**, 2528, 1990.
40. **Toti, F., Gachet, C., Ohlmann, P., Stierle, A., Grunebaum, L., Wiesel, M.-L., and Cazenave, J.-P.**, Electrophoretic studies on molecular defects of von Willebrand factor and platelet glycoprotein IIb-IIIa with antibodies produced in egg yolk from laying hens, *Haemostasis*, **22**, 32, 1992.
41. **Yoshimura, S., Watanabe, K., Suemizu, H., Onozawa, T., Mizoguchi, J., Tsuda, K., Hatta, H., and Moriuchi, T.**, Tissue specific expression of the plasma glutathione peroxidase gene in rat kidney, *J. Biochem.*, **109**, 918, 1991.
42. **Lee, K., Ametani, A., Shimizu, M., Hatta, H., Yamamoto, T., and Kaminogawa, S.**, Production and characterization of antihuman insulin antibodies in the hen's egg, *Agric. Biol. Chem.*, **55**, 2141, 1991.
43. **Morishita, E., Narita, H., Nishida, M., Kawashima, N., Yamagishi, K., Masuda, S., Nagao, M., Hatta, H., and Sasaki, R.**, Anti-Erythropoietin receptor monoclonal antibody: Epitope mapping, quantification of the soluble receptor, and detection of the solubilized transmembrane receptor and the receptor-expressing cells, *Blood,* **88**, 465, 1996.
44. **Stedman, R. A., Singleton, L., and Box, P. G.**, Purification of Newcastle disease virus antibody from the egg yolk of the hen, *J. Comp. Path.*, **79**, 507, 1969.
45. **Thalley, B. S. and Carroll, S. B.**, Rattlesnake and scorpion antivenoms from the egg yolks of immunized hens, *Biotechnol.*, **8**, 934, 1990.
46. **Bartz, C. R., Conklin, R. H., Tunstall, C. B., and Steele, J. H.**, Prevention of murine rotavirus infection with chicken egg yolk immunoglobulins, *J. Infect. Dis.*, **142**, 439, 1980.
47. **Yolken, R. H., Leister, F., Wee, S. B., Miskuff, R., and Vonderfecht, S.**, Antibodies to rotaviruses in chickens' eggs: A potential source of antiviral immunoglobulins suitable for human consumption, *Pediatrics*, **81**, 291, 1988.
48. **Ebina, T., Tukada, K., Umezu, K., Nose, M., Tsuda, K., Hatta, H., Kim, M., and Yamamoto, T.**, Gastroenteritis in suckling mice caused by human rotavirus can be prevented with egg yolk immunoblobulin (IgY) and treated with a protein-bound polysaccharide preparation (PSK), *Microbiol. Immunol.*, **34**, 617, 1990.
49. **Otake, S., Nishihara, Y., Makimura, M., Hatta, H., Kim, M., Yamamoto, T., and Hirasawa, M.**, Protection of rats against dental caries by passive immunization with hen-egg-yolk antibody (IgY), *J. Dent. Res.*, **70**, 162, 1991.
50. **Yokoyama, H., Peralta, R. C., Diaz, R., Sendo, S., Ikemori, Y., and Kodama, Y.**, Passive protective effect of chicken egg yolk immunoglobulins against experimental enterotoxigenic *Escherichia coli* infection in neonatal piglets, *Infect. Immun.*, **60**, 998, 1992.
51. **Gutierrez, M. A., Miyazaki, T., Hatta, H., and Kim, M.**, Protective properties of egg yolk IgY containing anti-*Edwardsiella tarda* antibody against paracolo disease in the

Japanese eel, Anguilla japonica Temminck & Schlegel, *J. Fish Dis.*, **16**, 113, 1993.
52. **Hatta, H., Mabe, K., Kim, M., Yamamoto, T., Gutierrez, M. A., and Miyazaki, T.,** Prevention of fish disease using egg yolk antibody, in *Egg uses and processing technologies-New developments*, Sim, J. S., and Nakai, S., Eds., CAB International, Oxon, 1994, p 241.
53. **Bishop, R. F., Davidson, G. P., Holmes, I. H., and Ruck, B. J.,** Virus particles in epithelial cells of duodenal mucosa from children with acute nonbacterial gastroenteritis, *Lancet*, **2**, 1281, 1973.
54. **Zoppi, G., Ferrarini, G., Rigolin, F., Bogaerts, H. and Andre, F. E.,** Response to RIT 4237 oral rotavirus vaccine in breast-fed and formula-fed infants, *Heiv. Peadiat. Acta*, **41**, 203, 1986.
55. **DeMol, P., Zissis, G., Butzler, J. P., Mutwewingabo, A. and Andre, F. E.,** Failure of live, attenuated oral rotavirus vaccine, *Lancet*, **2**, 108, 1986.
56. **Snodgrass, D. R. and Wells, P. W.,** Rotavirus infection in lambs: Studies on passive protection, *Arch. Virol.*, **52**, 201, 1976.
57. **Snodgrass, D. R. and Wells, P. W.,** The influence of colostrum on neonatal rotaviral infections, *Ann. Rech. Vet.*, **9**, 335, 1978.
58. **Offit, P. A., Shaw, R. D., and Greenberg, H. B.,** Passive protection against rotavirus-induced diarrhea by monoclonal antibodies to surface proteins vp3 and vp7, *J. Virol.*, **58**, 700, 1986.
59. **Ebina, T., Sato, A., Umezu, K., Ishida, N., Ohyama, S., Oizumi, A., Aikawa, K., Katagiri, S., Katsushima, N., Imai, A., Kitaoka, S., Suzuki, H., and Konno, T.,** Prevention of rotavirus infection by oral administration of cow colustrum containing antihuman rotavirus antibody, *Med. Microbiol. Immunol.*, **174**, 177, 1985.
60. **Loesche, W. J., Rowan, J., Straffon, L. H., and Loos, P. J.,** Association of *Streptococcus mutans* with human dental decay, *Infect. Immun.*, **11**, 1252, 1975.
61. **Bratthall, D. and Kohler, B.,** *Streptococcus mutans* serotypes: Some aspects of their identification, distribution, antigenic shifts, and relationship to caries, *J. Dent. Res.*, **55**, C15, 1976.
62. **Loesche, W. J. and Straffon, L. H.,** Longitudinal investigation of the role of *Streptococcus mutans* in human fissure decay, *Infect. Immun.*, **26**, 498, 1979.
63. **Lehner, T., Caldwell, J., and Smith, R.,** Local passive immunization by monoclonal antibodies against Streptococcal antigen I/II in the prevention of dental caries, *Infect. Immun.*, **50**, 796, 1985.
64. **Ma, J. K. C, Smith, R., and Lehner, T.,** Use of monoclonal antibodies in local passive immunization to prevent colonization of human teeth by *Streptococcus mutans*, *Infect. Immun.*, **55**, 1274, 1987.
65. **Filler, S. J., Gregory, R. L. , Michalek, S. M., Latz, J., and McGhee, J. R.,** Effect of immune bovine milk on *Streptococcus mutans* in human dental plaque, *Arch. Oral Biol.*, **36**, 41, 1991.

Chapter 12

MICROBIOLOGY OF EGGS

M. Kobayashi, M. A. Gutierrez, and H. Hatta

TABLE OF CONTENTS

I. Introduction
II. Bacterial Contamination and Protection of Eggs
    A. Egg Laying Process and Bacterial Contamination
        1. Bacterial Contamination in the Ovary and Its Prevention
        2. Bacterial Contamination in the Oviduct and Its Prevention
    B. Bacterial Contamination of Laid Eggs
        1. Microflora on the Surface of the Eggshell
        2. Role of Eggshell in Protection against Microbial Contamination
        3. Management of Eggs for Protection from Microbial Contamination
III. Eggs and *Salmonella*
    A. *Salmonella*-Associated Food Poisoning
        1. Current Status in Japan
        2. *Salmonella enteritidis* (SE)
        3. *Salmonella* Contamination
            a. In Egg Infection
            b. On Egg Infection
        4. *Salmonella* Multiplication in Shell Eggs
        5. *Salmonella* Multiplication in Liquid Eggs
    B. Commercialization of Eggs and *Salmonella* Contamination
    C. Prevention of *Salmonella* Contamination
        1. Pasteurization of Liquid Eggs
        2. Prevention of Stress
        3. Competitive Exclusion

References

I. INTRODUCTION

Hen eggs were once considered to be free of bacterial contamination. However, it is now known that even the surfaces of fresh eggs laid by healthy hens are usually contaminated by several kinds of bacteria. Eggs left at room temperature are sometimes spoiled by bacteria that penetrated into the eggs through their shells. The bacteria generally originate from excrement of the hen and the environment of the poultry farm. Recent studies have shown that certain pathogenic bacteria are present inside eggs laid by infected hens. *Salmonella* is one of those pathogenic bacteria often detected inside eggs. Needless to say, in order to protect the hen's embryo from bacterial contamination, eggs must have certain defense systems against invasion and multiplication of bacteria.

The first part of this chapter describes the bacterial contamination of eggs and protection of eggs from it, and the second part deals with several problems with special reference to *Salmonella enteritidis* (SE), which is considered to be a causative bacterium of food poisoning.

II. BACTERIAL CONTAMINATION AND PROTECTION OF EGGS

A. EGG LAYING PROCESS AND BACTERIAL CONTAMINATION

Before being laid, eggs can be contaminated by microorganisms via the mother's blood or via the ovary and oviduct, where pathogens can incorporate into the contents of the egg before the shell is completed. The controversy about transovarian transmission of pathogens have existed for a long time. Early studies by Gordon and Tucker sustained that bacteria can reach the ovaries through the blood [1]. To the contrary, Matthes and Hanschke affirmed that transovarian transmission by oral and intravenous routes does not occur [2].

As described in Chapter 1, the yolk follicles developed in the ovary are transferred one by one into the oviduct. Each of them is completed as an egg in every 25-27 hours. The ovary and oviduct are directly linked to the alimentary canal and the cloaca. Therefore, they run a great risk of being contaminated by migrating bacteria. These bacteria might, therefore, infect the egg contents during the egg formation process [3].

1. Bacterial contamination in the ovary and its prevention

Harry observed the existence of *Pasteurella haemolytica, Lactobacillus* sp., and *Micrococcus* sp. in the ovary and demonstrated that the ovary was not free from bacteria [4]. Even though some of the microorganisms isolated from the ovary could not be associated with the egg, he suggested that these bacteria may be transmitted through the ovary and multiply to cause spoilage of the egg. Keeping the ovary free from any bacterium seems to be impossible. However, as described in Chapter 11, egg yolk contains various antibodies (IgY) acquired from the blood of the mother hen. Therefore, it is highly likely that the IgY specific to bacteria contained in the egg yolk serve to inhibit or to depress the growth of bacteria that contaminated the egg through the ovary.

2. Bacterial contamination in the oviduct and its prevention

Egg yolk is at risk of bacterial contamination in the oviduct. But, bacteria are either eliminated or inactivated by the egg albumen proteins secreted from the albumen-secreting portion of the oviduct. It is known that egg albumen consists of various antimicrobial proteins, whose functional properties against bacterial multiplication are listed in Table 1. The thick albumen is formed in the uterus, and its high viscosity prevents the contaminating bacteria from migrating or spreading. Moreover, egg albumen contains IgM and IgA, which may contribute some resistance against invading bacteria.

The eggshell, the first barrier against bacterial penetration, is also formed in the uterus. After passing through the uterus, the egg is covered by the cuticle layer in the vagina and then, laid through the cloaca. Since the alimentary canal in birds joins the vagina just before the cloaca, the possibility for the eggshell to come in contact with feces is very high.

B. BACTERIAL CONTAMINATION OF LAID EGGS

Bacterial infection of eggs occurs mainly after laying. After oviposition, eggs are exposed to the contaminating organisms of the surrounding environment, feces, dust, and soil. It has been reported that several hundreds to millions of bacteria consisting of several species are generally observed on the surface of an eggshell [5]. The eggshell has about ten thousand pore canals with a diameter of 10-30 µm. This size is wide enough for bacteria to pass through. However, the eggshell is covered by a layer of cuticle on the outside and by the shell membranes on the inside. Their role in protection from bacterial invasion is described below.

Table 1
Antimicrobial Proteins in Egg Albumen

Protein	Function
Ovotransferrin	Chelates with Fe^{2+}, Cu^{2+}, Mn^{2+}, Zn^{2+}, depriving those elements needed for bacterial growth
Ovomucoid	Inhibits trypsin
Lysozyme	Hydrolyzes glyco moiety of peptidoglycans to kill gram positive bacteria
Ovoinhibitor	Inhibits trypsin, chymotrypsin, subtilisin, elastase
Flavoprotein	Binds riboflavin
Ovomacroglobulin	Inhibits trypsin, papain
Avidin	Binds biotin
Cystatin	Inhibits papain, ficin, bromelain

1. Microflora on the surface of the eggshell

Mayes and Takeballi investigated the bacterial species observed on the shell and reported that *Micrococcus* genera were the most common bacteria. *Achromobacter, Aerobacter, Alcaligenes, Arthrobacter, Bacillus, Escherichia, Flavobacterium, Pseudomonas,* and *Staphylococcus* were also found [6]. Although there was diversity in the bacterial species observed, bacteria originating from soil were relatively abundant.

In general, most of the bacteria found on the shell are gram positive, being the dominant *Micrococcus* genera which are strongly resistant to dryness. Gram negative bacteria are generally weak to dryness, so it may be difficult for them to grow on the eggshell surface. Even if both the gram positive and gram negative bacteria have the same probabilities of invading the egg contents through the shell, the multiplication of gram positive bacteria is prevented by lysozyme and other antibacterial proteins present in the egg albumen.

2. Role of the eggshell in protection against microbial contamination

The cuticle layer is the outermost layer of the egg. The cuticle, by covering the pore canals, prevents bacterial invasion. Changes in the cuticle structure influence the bacterial invasion after oviposition. Careless handling, such as deposition of feces, washing, and long storage times, damage the cuticle, and bacteria can invade the shell faster than before. Sparks and Board reported that the structure of cuticle changes depending on the moisture content of the air. Under wet conditions, the cuticle becomes a granular structure like foam, making eggs more susceptible to bacterial invasion [7]. Since bacteria require water to pass through the pore canal, it is important to keep the egg surface dry for prevention of microbial invasion.

The eggshell is another mechanical barrier against bacterial invasion. Shell thickness has a major effect on the ability of bacteria to find their way into the egg [8]. Damaged shells are less effective in preventing bacterial invasion. Abrasion, the amount of microorganisms, and the duration and extent of their contact with the shell have a direct influence on the shells capacity to prevent infection [9].

The shell membranes are situated on the inside of the eggshell. Their structure is like tangled threads made of protein and polysaccharides [10]. As early as 1940, Hains and Moran suggested that the role of the shell membranes was as bacterial filters [11]. The diameter of the opening space in the structure is around 1 μm, smaller than the pore canals on the eggshell, though the size is not small enough to prevent bacterial invasion completely. The shell membranes are the last structural barrier for defense against bacterial invasion.

3. Management of eggs for protection from microbial contamination

The shell of eggs is formed before the egg is laid through the cloaca. The alimentary canal connects to the oviduct just before the cloaca. Therefore, the first bacterial contamination on the eggshell occurs at this last part of the oviduct, because this part is usually polluted by feces. It is known that a great number of intestinal microorganisms are present in hen excrement, and the microflora of feces are influenced by both health conditions and feed. In order to reduce the amount of bacteria in feces, especially those which cause food poisoning, poultry farms should provide hens with a sanitary, less-stressful environment. It is absolutely necessary not to use feed or water contaminated with pathogenic bacteria.

In poultry houses, laid eggs are exposed to the possibility of secondary contamination via the excrement adhering to the cages or the floating bacteria caused by air conditioning, etc. This kind of bacterial contamination can easily be avoided by correct farm management. Periodical washing of poultry facilities, sterilization, installation of air filters, etc., are essential to decrease the probability of secondary bacterial contamination.

The eggshells are generally washed to remove all dirt, including excrement adhering to the shell. However, washing eggs removes the cuticle layer at the same time, enhancing the opportunity of bacterial invasion into the egg. On washing eggs, the water temperature should be controlled to avoid bacterial contamination. If the temperature of the washing liquid is lower than that of eggs, the bacteria present inside the pore canals are sucked into the egg. Bean and MacLaury have explained that the absorption of bacteria into the interior of eggs is induced by shrinking of the shell membranes. When the shrinking occurs, the membranes are peeled from the inner surface of the shell, generating a vacuum effect that brings the bacteria into the egg. In fact, this is a consequence of the temperature difference between the outside and the inside of the egg [12]. Therefore, the water for washing should be more than 30°C, followed by fast drying of the washed eggs and storage at a temperature of 5°C or slightly lower [13].

III. EGGS AND *SALMONELLA*

A. *SALMONELLA* ASSOCIATED FOOD POISONING
1. Current status in Japan

The majority of food poisoning cases in Japan were once a consequence of the long traditional custom of eating raw seafood. This kind of food poisoning was caused in most cases by gastroenteritis associated with *Vibrio* sp. However, around 1989, a sudden increase in outbreaks of *Salmonella*-associated food poisoning was observed. In 1992, 144 cases involving 11,431 patients were reported. This number of cases is higher than that of *Vibrio*-associated food poisoning [14]. This tendency still continues, and there are no signs of decrease in the number of cases of *Salmonella* food poisoning. It was once thought that the food products susceptible to *Salmonella* contamination were all kinds of meat and meat products. However, since 1989, hen eggs have been involved in many cases of food poisoning [15, 16]. From 1989 to 1993, 181 outbreaks of *Salmonella* poisoning were reported, 59 of them confirmed to be foodborne outbreaks. Egg omelets, kinshi eggs (crepe eggs), mayonnaise, tiramisu, and other egg-containing foods accounted for 70% of those 59 outbreaks [17].

Salmonella is a motile, gram negative bacillus, which has been classified into two serotypes: O antigen and H antigen. Of the 2,000 strains of *Salmonella*, about 100 are known to produce food poisoning in humans. The symptoms of gastroenteritis caused by association with *Salmonella* develop in 8 to 24 h after oral infection. These symptoms include diarrhea, abdominal pain, chills, fever, vomiting, headache, and acute inflammation of the stomach and intestines. Several species of *Salmonella* have been isolated from eggs. The most common are *S. typhimurium, S. enteritidis, S. infantis, S. thompson, S. montevideo, S. litchfield*, which are disease causative species. Although they are pathogenic to humans, they produce minor mortality in hens. Therefore, some hens infected with these kinds of *Salmonella* maintain their normal egg-laying rate, transferring the microbe to their eggs. *S. pullorum* also causes food poisoning, and its outbreaks are very critical for the poultry industry because the disease brings very high mortality in hens, producing important economic losses [18].

2. *Salmonella enteritidis* (SE)

In recent years, the rate of SE as the origin of *Salmonella* food poisoning has increased sharply in Japan. The same tendency is occurring in the United States, Europe, and many other countries. In the five years between 1984 and 1988, 70 outbreaks of *Salmonella* food poisoning were reported. SE accounted for only 3 of them. *S. typhimurium* and *S. litchfield* accounted for most of those cases [19]. However, in 1989, SE was involved in 23 of 33 outbreaks, and this increasing tendency continues [20].

The species of *Salmonella* have been classified according to their serotype. From the food hygiene point of view, *Salmonella* is seriously dangerous. Therefore, a classification based on serotype may not be enough. Strain differences must be taken into consideration when countermeasure methods to control this bacteria are planned. A classification based on phage types is now generally used. The SE strains detected in Japan have been classified into phage types 4, 5, 8, and 34 (phage type 34 is found only in Japan) [16].

According to annual data, the occurrence of SE of different phage types fluctuated before 1988, when phage type 8 accounted for 60% of the cases. In 1989, phage type 34 appeared, and since then, the occurrence rate has changed. In 1991, phage type 34 accounted for 33.5% of the outbreaks while phage type 8 was only 3.5% [17].

3. *Salmonella* contamination
a. In egg infection

SE and *S. typhymurium* are species that have the capacity of establishing themselves inside the ovaries. Nakamura reported that SE and *S. typhymurium* are transferred directly from the ovary into the egg yolk before the shell is formed [21]. This vertical (transovarian) transfer of bacteria is called "In egg infection".

In the United Kingdom, to demonstrate the in egg infection, SE phage type 4 was directly inoculated into egg laying hens (10^2-10^6 cfu/hen). Isolation of the bacterium from the laid eggs was then performed. In addition, the hens were dissected and all the internal organs were inspected. The bacterium was isolated from 25% of all livers, spleens, ovaries, and oviducts analyzed. In eggs, the ratio of bacterial detection was 1.1% (4 out of 375) [22]. In another experiment, SE was inoculated into hens at a dose of 10^6 cfu/hen. The rate of bacterium isolated from eggs was 5% in the 2nd day postinoculation (PI), 4.5% in the 4th day PI, and 15% in the 13th day PI. The bacterium was detected from internal organs even 42 days PI [23]. This fact demonstrates how difficult it is to clear bacteria from the hen's body after the infection is established. The complete extermination of *Salmonella* from infected hens is very difficult.

In eggs laid by SE-carrying hens, a very high percentage of infection could be expected. However, the infection rate is very low. Humphrey and his colleagues reported that the bacterium was positively detected in only 11 of 1,119 (1%) eggs laid by 35 infected hens [22].

In another investigation made in a poultry house suspected to be contaminated with SE, only 0.6% of the produced eggs (32 out of 5,700) were found to be infected [24]. Therefore, the highest probability of contamination of eggs laid by SE-carrying hens can be estimated to be less than 1%. In Japan, an investigation carried out on contamination of eggs packed for marketing from 1990 to 1991 reported the presence of *Salmonella* in 0.09% of the eggs sampled (6 out of 6, 700 eggs) [25].

b. On egg infection

Except for SE and *S. typhymurium*, other *Salmonella* bacteria show a weak invasive ability and their establishment in the ovary is almost negligible [26-28]. Considering the way for *Salmonella* to infect eggs, "on egg infection" seems to occur more frequently than "in egg infection." In the case of "on egg infection," the shells are contaminated by feces containing *Salmonella*. Then, the microbe penetrates into the eggs through the shell [29].

In order to investigate "on egg infection," SE was applied directly over the eggshell, kept at 25°C and then, the time for penetration into the egg was determined [30]. After 9 h incubation, the bacterium penetrated into the egg albumen, and 24 h later, SE reached the egg yolk. A similar experiment was also performed with eggs having normal or cracked shells. Both groups of eggs were immersed in a SE suspension at 25°C. Compared with normal eggs, the number of bacteria invading the cracked eggshells was remarkably higher [25]. It was also observed that the number of invading bacterial cells was higher in washed eggs than in unwashed eggs. Washing removed the cuticle from the shell surface making it easier for SE to invade the eggs [25].

4. *Salmonella* multiplication in shell eggs

To investigate the way of multiplication of *Salmonella* inside shell eggs, 2 groups of eggs were inoculated with SE (65 cfu) each. One group was kept at 30°C and resulted in bacterial multiplication of levels of more than 10^7 cfu/ml within 24 h. In the other group of eggs, which was kept at 4°C, no bacterial proliferation was observed. However, the bacterium inoculated was still alive even after 11 days of storage [31]. In another experiment with quail eggs, a small hole was made on the shell and 10 cfu of SE was inoculated. No increase of bacterium was observed at 4°C. However, the bacterium multiplied to 10^2 cfu/ml after 20 days at 10°C. In the same experiment, *S. typhimurium* was inoculated in the same way as above. The experimental temperatures were 4°C and 15°C. No multiplication occurred at 4°C. On the other hand, the bacterium reached levels of 10^8 cfu/ml after 3 days at 15°C. These results indicate a big difference in the rate of multiplication of *Salmonella* at temperatures of 10°C and 15°C [32].

It has been reported that the pH range for optimum multiplication of *Salmonella* is from 4.5 to 9.0. In the case of "on egg infection," even if *Salmonella* passed through the eggshell, alkaline pH and the presence of lysozyme, conalbumin, and other biologically active proteins prevent a sudden multiplication of the bacterium. Thus, the numbers of bacteria reaching the egg yolk might not be considerable. In the case of "in egg infection," the egg yolk membrane has already been contaminated. The contaminating bacteria multiply easily depending on the temperature.

5. *Salmonella* multiplication in liquid eggs

In liquid eggs, the bacterial multiplication pattern is different from that of shell eggs. In an investigation carried out on the multiplication of *Salmonella* (*S. typhimurium, S. cerro, S. infantis*), they were inoculated into liquid whole eggs, liquid egg albumen, and liquid egg yolk at concentrations of 10^2-10^3 cfu/g, and then, stored at 5°C. No propagation of bacteria was observed after standing for 8 days. At 25°C, the liquid whole egg and liquid egg yolk produced bacterial propagation that reached levels of 10^8-10^9 cfu/g, after standing for only 24 h. However, in egg albumen, after 4 days, the bacteria increased only one order [33].

Experiments to make clear the survival capacity and stability of the bacteria showed that low temperatures prevent bacteria growth. However, *Salmonella* can survive even in frozen liquid eggs [34]. Therefore, in order to prevent a sudden multiplication of *Salmonella*, it is important to consider the temperature to preserve liquid eggs, especially yolk containing liquid eggs.

B. COMMERCIALIZATION OF EGGS AND *SALMONELLA* CONTAMINATION

In 1994, the total production of hen eggs in Japan was 2,480,000 tons (including shell), 14% of which were processed as liquid eggs. The ratio of *Salmonella* detection in shell eggs on the market is, as mentioned above, only around 0.1%. On the other hand, many reports about *Salmonella* detection revealed a bacterial contamination ratio considerably higher in liquid eggs than in shell eggs. This fact indicates that the bacterial contamination occurs during the manufacturing processes.

It has been reported in Australia, that the level of bacterial detection from equipment for the production of liquid eggs (bulk tank, etc.) was 15.3%, while the level found in shell eggs used for liquid eggs is only 0.2% [35]. Several researchers reported that at certain plants in Japan, *Salmonella* contamination was observed in liquid eggs, though no *Salmonella* was detected in at least 1,500 shell eggs used as raw material [36].

In the liquid egg production process, bacterial dissemination is highly probable. For example, around 20,000 eggs (58-64 g of M size) are required to produce 1 ton of liquid whole eggs. If, among these eggs, there are two eggs contaminated with 10^5 *Salmonella* cells each, there would be an average of 20 bacterial cells for every 100 g of liquid whole eggs produced.

C. PREVENTION OF *SALMONELLA* CONTAMINATION
1. Pasteurization of liquid eggs

Shell eggs can not be pasteurized, but liquid eggs have the advantage that they can be pasteurized by a heat treatment. Pasteurization of liquid eggs is usually done in a plate type or cylinder type heat exchanger at temperatures high enough to kill *Salmonella* but without solidification of the egg.

There have been many papers reporting the relationship between the heat resistance of *Salmonella* and the effects of concentration of protein, lipids, and the addition of salt or sugar. The heat resistance of bacteria in egg albumen, egg yolk, and in the whole egg was investigated at 60°C. The D value (required time in min. to decrease the surviving bacterial cells to one tenth) was 0.2 min. and 0.4 min. for egg albumen and the whole egg, respectively, while that for yolk was 1.1 min [37]. Bacterial heat resistance is also greatly influenced by pH. In general, the heat resistance of *Salmonella* is relatively high at pH 5-6, but under alkaline conditions, it is reduced. For instance, if the pH value of liquid whole egg is raised by one unit, the D value drops to less than a half [38, 39]. Both sugar and salt increase the heat resistance of *Salmonella*. This means that egg product containing sugar or salt requires higher temperature or longer heating times to be well pasteurized. Ng and his co-workers tested 300 *Salmonella* isolates and found that the heat resistance ranged from two-thirds to two times the resistance of *S. typhimurium* TM-1 (2.2 min. at 60°C) [40]. Therefore, in order to get an adequate pasteurization effectiveness, the temperatures and times of the heating treatment should be sufficient to give adequate kills of *Salmonella* strains that have two times the D value of *S. typhimurium* TM-1, under any conditions of pH, temperature, and salt or sugar content.

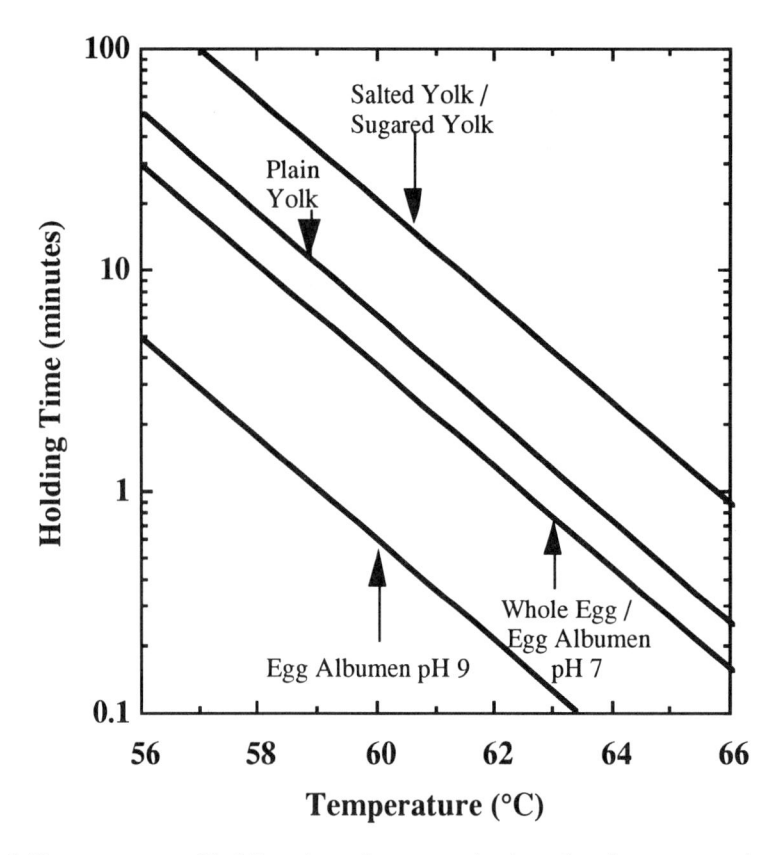

Figure 1. Temperature and holding times for pasteurization of various egg products. Lines show the necessary conditions to obtain adequate kills of *Salmonella* strains that have two times the D value of *S. typhimurium* TM-1.

Figure 1 shows the known relationship between time and temperature required to produce equal pasteurization effectiveness in various egg products. At 60°C, egg albumen (pH 9.0) can be pasteurized in 0.6 min and whole egg and egg albumen (pH 7.0) in 3.5 min. Plain yolks need 6 min. while salted or sugared yolks will be equally pasteurized after 20 min. In the United States and Japan, 9 D value (required time in min. to decrease surviving *Salmonella* cells to $1/10^9$) used for pasteurization is based on the 9 D value of *S. typhimurium* Tm-1, which is relatively resistant to heat (9 D value, 3.5 min, at 60°C) [41]. Under these conditions, any strain of *Salmonella* can be eradicated. FAO/WHO, in their international recommendations for management of the microorganisms of hen egg and its products, recommends the pasteurization at 64°C for 2.5 min. [42]. In many countries of Europe, the pasteurization conditions are more strict than in Japan and the United States. Table 2 shows a list of pasteurization conditions (temperature and time) used in various countries.

2. Prevention of stress

Salmonella is detected very frequently in the intestines, especially in the appendix caecum of experimentally infected hens, and its presence in the feces persists for up to 6 weeks [29]. Therefore, all eggs laid in this period are considered to be under risk of contamination. There are many interesting data regarding the relationship between stress of hens and increase in *Salmonella* cellular counts in the excreted feces. In the United States, poultry reared at high temperatures showed an increased number of excreted bacteria [43]. Other authors also reported the relationship between the increase of excreted *Salmonella* and stress produced by cold or problems with rearing facilities, like interruption in water supply, etc. [44-46]. When the egg laying rate of some hens decrease, feeding interruption is a general method to improve the laying rate and the quality of the eggshell.. However, this procedure causes great stress on those hens and increases the cell counts of excreted *Salmonella* [47]. Methods to improve the egg laying rate usually result in increased *Salmonella* due to "on egg infection." Therefore, in order to avoid *Salmonella* contamination, it is necessary to manage farm conditions to avoid exposing hens to any kind of stress [48].

Table 2

Conditions of Pasteurization for Liquid Eggs

	Egg White	Egg Yolk	Whole Egg
U.S.A.	56.7 °C, 3.5 min	61.1°C, 3.5 min	60.0°C, 3.5 min
Japan	55-56°C, 3.5 min	60.0°C, 3.5 min	60.0°C, 3.5 min
U. K.	57.2°C, 2.5 min	64.4°C, 2.5 min	64.4°C, 2.5 min
Germany	56.0°C, 8.0 min	58.0°C, 3.5 min	65.5°C, 5.0 min
France	55.5°C, 3.5 min	62.5°C, 4.0 min	58.0°C, 4.0 min
Denmark	61.0°C, 3.0 min	68.0°C, 4.5 min	68.0°C, 4.5 min

3. Competitive exclusion

The continuous increase of food poisoning outbreaks related to SE infected chickens has become a worldwide problem. As a countermeasure to eliminate *Salmonella* from poultry farms, the World Health Organization (WHO) prescribed the following three approaches in 1989 [49]:

1) Washing and disinfection of the chicken farm, separation of infected chickens, and extermination of rats.

2) Appropriate monitoring system to inspect for *Salmonella*.

3) Application of vaccines, antibacterial drugs, and the competitive exclusion (CE) method.

Among these methods, the CE method consists of feeding newly born chicks with fresh excrement from adult hens. In this way, various other bacteria will occupy the young chick's intestine. Even if oral infection with *Salmonella* occurs, the bacteria will not be able to multiply and will be removed from the intestine in a short time [15, 50]. The intestine of newly born chicks is almost free of bacteria, and at this time, the sensitivity to *Salmonella* is extremely high. The bacterium will be easily established even by small numbers of *Salmonella* cells if orally administered [53]. The normal bacteria flora in adult hen's intestine are mainly anaerobic species. These bacterial flora multiply rapidly and cover the intestinal mucosa of young chick's intestine. Under this condition, *Salmonella* is not able to fix on the intestinal wall. Researches and field tests with the CE method have been performed in Europe for the last 20 years and showed that the CE method is effective in decreasing *Salmonella*-carrying hens [49, 51].

Furthermore, the method is very easy to apply because newly born chicks are fed only one time with CE-culture products. In addition, compared with vaccination and/or drug administration, CE has no risk of chemical residues neither in the egg nor in the poultry meat. When young chicks are still in the incubator, this method is very easy and effective for controlling *Salmonella* contamination. In experiments done in incubators, contamination by *Salmonella* were found in the intestines of chickens as young as 10 days old [52]. Therefore, the effective countermeasure against the bacterial contamination should be applied before that time.

In Finland, anaerobic culture products of intestinal appendix contents from adult hens are being marketed. The products are used in approximately 90% of broiler poultry farms. The usage of this culture is spreading to the United Kingdom and many other countries of northern Europe [49]. Since this excrement culture includes a great diversity of grown bacteria, which are very difficult to identify, the name "undefined culture" is used to characterize it. In past experience on the application of CE, neither an increase in mortality or damage to egg productivity have been reported so far. However, it is difficult to consider this method as completely safe, because there may be the possibility that the undefined culture contains some unknown pathogenic agent [53].

In the United States, trials with undefined culture have started, but since the effect is not clear, permission from the Federal Drug Administration (FDA) for marketing it is under suspension. However, under the recommendation of the WHO and because CE is being used worldwide, the Agricultural Research Service (ARS) of the United States Department of Agriculture (USDA), has started research to develop a "defined culture." Investigation of the role of sugars in interference with *Salmonella* fixation found that, when lactose was metabolized by intestinal flora, a clear obstruction to *Salmonella* fixation occurred. In the intestinal appendix, certain bacteria (except *Salmonella*) ferment lactose to produce lactic acid which is converted to propionic and other acids by *Veillonella* and other bacteria [53]. These fermentation products might play an important role in interfering with *Salmonella* fixation, because in newborn chicks treated with CE, the amount of propionic acid was 10-20 times more than that without CE treatment [53].

Based on these results, the ARS developed the continuous flow culture system to obtain a defined culture for controlling *Salmonella* spreading. This culture, named CF-III, has demonstrated clear interference with *Salmonella* fixation in laboratory scale experiments as well as in large scale field trials [53]. However, *Salmonella* is not always eliminated by this method alone. A combination of CE with the general method of disinfection, extermination of rats, and application of vaccines is necessary to eradicate *Salmonella* completely.

REFERENCES

1. **Gordon, R. F. and Tucker, J. F.**, The epizootiology of *Salmonella menston* infection of fowls and the effect of feeding poultry food artificially infected with *Salmonella*, *Br. Poult. Sci.*, **6**, 251, 1965.
2. **Matthes, S. and Hanschke, J.**, Experimentelle untersuchungen zur übertragung von bakterien über das Hühneria, *Berl. Muench. Tieraerztl. Wochenschr.*, **90**, 200, 1977.
3. **Board, R. G., Clay, C., Lock, J., and Dolman, J.**, The egg: A compartmentalized, aseptically packaged food, in *Microbiology of the Avian Egg*, Board, R. G. and Fuller, R., Eds., Chapman & Hall Inc., London, 1994, p 43.
4. **Harry, E. G.**, Some observation on the bacteria content of the ovary and oviduct of the

fowl, *Brit. Poult. Sci*, **4**, 63, 1963.
5. **Bruce, J. and Drysdale, E. M.**, Trans-shell transmission, in *Microbiology of the Avian Egg*, Board, R. G. and Fuller, R., Eds., Chapman & Hall Inc., London, 1994, p 63.
6. **Mayes, F. J. and Takeballi, M. A.**, Microbial contamination of the hen's egg: A review, *J. Food Protect.*, **46**, 1092, 1983.
7. **Sparks, N. H. C. and Board, R. G.**, Bacterial penetration of the recently oviposited shell of hen's eggs, *Aust. Vet. J.*, **6**, 169, 1985.
8. **Sauter, E. A. and Peterson, C. F.**, The effect of egg shell quality on penetration by various *Salmonellae*, *Poult. Sci.*, **53**, 2159, 1974.
9. **Burley, R. W. and Vadehra, D. V.**, The microbiology of avian eggs, in *The Avian Egg, Chemistry and Biology*, Burley, R. W. and Vadehra, D. V., Eds., John Wiley & Sons, New York, 1989, p 289.
10. **Baker, J. R. and Balch, D. A.**, A study of the organic material of hen's egg shell, *Biochem. J.*, **82**, 352, 1962.
11. **Haines, R. B. and Moran, T.**, Porosity of and bacterial invasion through the shell of hen's egg, *J. Hyg.*, **37**, 453, 1940.
12. **Bean, K. C. and MacLaury, D. W.**, The bacterial contamination of hatching eggs and method for its control, *Poult. Sci.*, **38**, 693, 1959.
13. **Watanabe, A.**, The condition of penetration of *Salmonella* through the egg shell, *New Food Industry (in Japanese)*, **37**, 68, 1995.
14. **Ito, T.**, 4 th Seminar to exterminate salmonellosis, Tokyo, Japan, Keiranniku Joho Center, 1995, (in Japanese).
15. **Chazono, A.**, The contamination and countermeasure of salmonellosis in the world, *J. Jap. Soc. Poult. (in Japanese)*, **30**, 72, 1993.
16. **Katsube, Y.**, Environmental contamination of *Salmonella*, *Nyugikyo Shiryo (Bull. Jap. Dairy Tech. Assoc.) (in Japanese)*, **39**, 89, 1989.
17. **Ito, T.**, 2nd Seminar to exterminate salmonellosis in Tokyo, Japan, Keiranniku Joho Center, 1994, (in Japanese).
18. **Nakamura, M.**, The contamination and coutermeasure of salmonellosis in poultry, *Chikusan no Kenkyu (Animal Husbandry) (in Japanese)*, **44**, 233, 1990.
19. **Kusunoki, J. and Ota, K.**, Outbreaks of food poisoning by *Salmonella enteritidis*, *Monthly Epidemiological Record, Tokyo (in Japanese)*, **10**, 1, 1989.
20. **Sato, S.**, The contamination and countermeasure of salmonellosis in egg, *Yokei no Tomo (in Japanese)*, **340**, 10, 1990.
21. **Nakamura, M.**, Report of WHO consultation on epidemiological emergency in poultry and egg salmonellosis (1), *Keibyo Kenkyu Kaiho (Bull. of Jap. Soc. on Poult. Dis.) (in Japanese)*, **25**, 127, 1989.
22. **Humphrey, T. J., Baskerville, A., Chart, H., and Rowe, B.**, Infection of egg-laying hens with *Salmonella enteritidis* by oral inoculation, *Vet. Rec.*, **125**, 531, 1989.
23. **Timoney, J. F., Shivaprasad, H. L., Baker, R. C., and Rowe, B.**, Egg transmission after infection of hens with *Salmonella enteritidis* phage type 4, *Vet. Rec.*, **125**, 600, 1989.
24. **Humphrey, T. J., Whitehead, A., Gawler, A. H. L., Henley, A., and Rowe, B.**, Numbers of *Salmonella enteritidis* in the contents of naturally contaminated hen's eggs, *Epidemiol. Infect.*, **106**, 489, 1991.
25. **Konuma, H. and Shinagawa, K.**, Microbiological sanitary control for egg and egg products. Grading and packaging center of fresh egg, *Syokuhin Eisei Kenkyu (Food Sanitation Res.) (in Japanese)*, **43**, 49, 1993.
26. **Hopper, S. A. and Mawer, S. L.**, *Salmonella enteritidis* in a commercial layer flock, *Vet. Rec.*, **123**, 351, 1988.

27. **Gast, R. K. and Beard, C. W.,** Isolation of *Salmonella enteritidis* from internal organs of experimentally infected hens, *Avian Dis.*, **34**, 991, 1990.
28. **Gast, R. K. and Beard, C. W.,** Production of *Salmonella enteritidis*-contaminated eggs by experimentally infected hens, *Avian Dis.*, **34**, 438, 1990.
29. **Shizaprasad, H. L., Timoney, J. F., Morales, S., Lucio, B., and Baker, R. C.,** Pathogenesis of *Salmonella enteritidis* infection in laying chickens, 1. Studies on egg transmision, clinical signs, fecal shedding, and serologic responses, *Avian Dis.*, **34**, 548, 1990.
30. **Ohnaka, T., Nakajima, K., Sugimoto, M., Kimura, S., Kitano, Y., Yoshino, M., and Shimosakko, H.,** Bacteriological survey of liquid egg, *Jap. J. Food Microbiol. (in Japanese)*, **9**, 109, 1992.
31. **Kobayashi, K., Harada, K., Araya, H., and Sasabe, Y.,** Growth and survival of *Salmonellae* in eggs and egg products, *Bull. Osaka Pref. Inst. Publ. Hlth. (in Japanese)*, **21**, 21, 1983.
32. **Ito, T., Takahashi, M., Saito, K., Inaba, M., Kai, A., Yanagawa, Y., Takano, I., Sakai, S., Shinohara, T., Tamura, N., Kato, K., Tsuchiya, N., Kinoshita, M., Nakajima, M., Takatori, S., and Tazaki, T.,** An outbreak due to *Salmonella typhimurium*, biotype 17i, associated with the use of raw quail eggs and incidence of *Salmonella* in quail eggs, *Ann. Rep. Tokyo Metr. Res. Lab. P.H. (in Japanese)*, **34**, 126, 1983.
33. **Shiozawa, K., Sugieda, M., Hayashi, M., Handa, Y., Nishina, T., Nakatsugawa, S., and Akahori, K.,** Microbial contamination of liquid egg and the growth of *Salmonella* and pathogenic *Escherichia coli* in liquid egg, *Bull. Shizuoka Pref. Inst. Publ. Hlth. Environ. Sci. (in Japanese)*, **30**, 61, 1987.
34. **Harada, T., Watanabe, H., Shirase, I., Wada, A., Nagasawa, S., Orinara, N., Saishoji, Y., Yonemochi, M., and Kaiho, Y.,** Contamination, heat resistance and survival in storage of *Salmonella* in liquid eggs, *Food Sanitaton Res. (in Japanese)*, **43**, 47, 1993.
35. **Peel, B.,** Occurence of *Salmonella* in raw and pasteurized liquid whole egg, *Queensland J. Agric. and Anim. Sci.*, **33**, 13, 1976.
36. **Suzuki, A.,** Powdered egg white, frozen liquid egg white and *Salmonella*, *Modern Media (in Japanese)*, **12**, 24, 1966.
37. **Humphrey, T. J., Chapman, P. A., Rowe, B., and Gilbert, R. J.,** A comparative study of the heat resistance of *Salmonella* in homogenised whole egg, egg yolk or albumen, *Epidemiol. Infect.*, **104**, 237, 1990.
38. **Cotterill, O. J.,** Equivalent pateurization temperature to kill *Salmonellae* in liquid egg white at various pH levels, *Poult. Sci.*, **47**, 354, 1968.
39. **Anellis, A., Lubas, J., and Rayman, M. M.,** Heat resistance in liquid eggs of some strains of the genus *Salmonella*, *Food Res.*, **19**, 377, 1954.
40. **Ng, H.,** Heat sensitivity of 300 *Salmonella* isolates. In *The Destruction of Salmonellae, a Report of the Western Experiment Station Collaborators Conference*, U.S. Dept. Agric., Agric. Res. Serv., 1996.
41. **Garibaldi, J. A., Straka, R. P., and Ijichi, K.,** Heat resistance of *Salmonella* in various egg products, *Appl. Microbiol.*, **17**, 491, 1969.
42. **FAO/WHO,** Proposed draft code of hygienic practice for egg products, July, 1974.
43. **Jones, F. T.,** Breeder flock study shows *Salmonella*-causing factors, *Feedstuffs*, **64**, 1, 1992.
44. **Soerjadi, A. S., Druitt, J. H., Lloyd, A. B., and Cumming, R. B.,** Effect of environmental temperature on susceptibility of young chickens to *Salmonella typhimurium*, *Aust. Vet. J.*, **55**, 413, 1979.
45. **Weinack, O. M., Snoeyenbos, A. S., Soerjadi-Liem, A. S., and Smyser, C. F.,** Influence

of temperature, social, and dietary stress on development and stability of protective microflora in chicken against *S. typhimurium*, *Avian Dis.*, **29**, 1177, 1985.
46. **Brownell, J. R., Sadler, W. W., and Fanelli, M. J.**, Factors influencing the intestinal infection of chickens with *Salmonella typhimurium*, *Avian Dis.*, **13**, 804, 1969.
47. **Holt, P. S. and Porter, R. E. J.**, Effect of induced molting on the recurrence of a previous *Salmonella enteritidis* infection, *Poult. Sci.*, **72**, 2069, 1993.
48. **Nakamura, M.**, The effect of stress on *Salmonella* infection in chickens, *Bull. of Jap. Soc. Poult. Dis. (in Japanese)*, **29**, 136, 1993.
49. **Sato, S.**, 4th Seminar to exterminate salmonellosis, Tokyo, Keiranniku Joho Center, 1995, (in Japanese).
50. **Fukuda, T.**, 4th Seminar to exterminate salmonellosis, Tokyo, Keiranniku Joho Center, 1995, (in Japanese).
51. **Nurmi, E., Nuotio, L., and Schneitz, C.**, The competitive exclusion concept: Development and future, *Int. J. Food Microbiol.*, **15**, 237, 1992.
52. **Stavric, S., Gleeson, T. M., Blanchfield, B., and Pivnick, H.**, Competitive exclusion of *Salmonella* from newly hatched chicks by mixtures of pure bacterial cultures isolated from fecal and caecal contents of adult birds, *J. Food Prot.*, **48**, 778, 1985.
53. **John, R. D.**, 4th Seminar to exterminate salmonellosis, Tokyo, Keiranniku Joho Center, 1995, (in Japanese).

INDEX

A

N-acetylglucosamine	20, 21, 45, 46
β-N-acetylglucosaminidase	39, 136
parameter of pasteurization	139
N-acetylmuramic acid	45, 46
N-acetylneuraminic acid	20-22, 105, 106
acid phosphatase	136-138
acid proteinase	135, 136
acid value	80
acquired immunity	152
adenovirus	
diagnosis using IgY	166
air chamber	
location in egg	2
albumen	
external thin albumen	19
internal thin albumen	19
thick albumen	19
role in bacterial prevention	180
albumen secreting portion	9, 10
alkaline phosphatase	136-138
allergy	
induced by egg protein	31
aluminum sulfate	
pasteurization of egg	127
alzheimer's disease	87
amino acids	22, 57, 64
essential amino acids	26, 27
p-aminobenzoic ethyl ester (ABEE)	107
aminocarbonyl reaction	22
aminopeptidase	135, 137, 138
glutamylaminopeptidase	137
methionine-preferring aminopeptidase	137
amylase	136
in livetin fraction	17
parameter of pasteurization	141
anterior lobe of pituitary	
effect on hen ovulation	8
antimicrobial proteins	
functional properties	181
antibodies (*see* IgY)	
IgA and IgM in egg	153, 180
antioxidant activity	
egg yolk proteins	66
egg yolk lipids	90
arachidonic acid (AA)	17, 78, 85, 87, 88, 93
arginine vasotocin (AVT)	
in oviposition process	10
arsenic	80
arthritic patients	92
atopy	31
ATPase	136, 139
avidin	19, 22, 31, 39, 181
supressing cell growth	148

B

bacteria	
gram negative	181
gram positive	181
phagetype	182
serotype	183
bacterial contamination	
in the ovary	180
in the oviduct	180
of laid eggs	180
prevention of	180, 181
biotin	22, 31, 65
bovine serum albumin	
emulsifying ability of	131

C

carbohydrates	
content in egg	14
content in egg yolk	18
content in egg white	22
nutritional aspect	28
of cuticle	2
oligosaccharide	28
sialic acid	60, 103, 148
sialyloligosaccharide	28
cardiovascular disease	32
carbonyl value (CV)	66, 80

carotene
    α-, and β-carotenes   18
carotenoids
    antitumor activity   28
    content in egg yolk   18
λ-carrageenan
    purification method of IgY   159, 160
calcite crystals
    in eggshell   15
catalase   136
    parameter of pasteurization   141
cathepsins
    yolk-cathepsin D1 and D2   136
CD analysis
    heat denatured OVA   126
cell growth promoting activity
    hen egg components   146
cell proliferation promoting factors
    in hen egg   145
    in egg yolk   147
    in egg white   148
cerebrosides
    in egg yolk   18
chalaza   105
chalazaecode
    function of   6
    location in egg   2
chalaziferous layer
    binding with lysozyme   21
    formation of   10
    location in egg   2
    percentage in egg albumen   19
    structure of   6
chicken serum albumin   158
chitin   21
cholesterol   79, 85, 90
    content in egg   31
    content in egg yolk   18, 67
    content in EYP-80   67
cholesterol ester
    content in egg yolk   18
choline   65, 86-87
cholinesterase   136, 141
    in livetin fraction   17
circular dichroism (CD)   126
cis-diammine dichloroplatinum   92
cloaca
    location in oviduct   9, 10

collagen
    collagen-like protein
        in shell membrane   5
    in the shell   4
    type I, type V,
        and type X collagens   5
competitive exclusion (CE)
    CE-method   187, 188
    defined culture   188
complement
    association with IgY   161
conalbumin (see ovotransferrin)
    promoting cell growth   146, 148
corn oil
    hens diet supplemented with   30
coronary heart disease   31
coronavirus   109
cryptoxanthin
    quantities in egg yolk   18
cuticle
    amino acid composition of   4
    component of   2, 15
    effect of washing   182
    location in eggshell   3
    role in bacterial prevention   181
    structural change of   181
cystatin
    (see ficin-papain inhibitor)   19, 22, 39

D

deaminase
    against nuclosides   136, 140
deoxyribonuclease   136
desmosine
    in shell membrane   5
diarrhea   110
1,25-dihydroxyvitamin D
    application of IgY   166
dementia   87
docosahexaenoic acid (DHA)   17, 32, 78, 87-91, 93
D value   185

E

Edwardsiella tarda
    Edwardsiellosis of
        Japanese eels   171

egg albumen
  (*see* egg white protein)
    amino acid composition 26
    content of ovomucin 6
    denaturation temperature 119
    foaming ability 126
    foaming stability 126, 128
    gelation and
      agglutination of 119
    role in bacterial prevention 180
    stored in oviduct 9, 10
    thick and thin albumen 6
    viscosity 127, 128
egg proteins 37, 38
    characteristic properties of 118
    distribution in egg 14
    essential amino acids 26, 27
    in embryogenesis 37, 38, 40, 52
    N-linked sugars 37, 38
    nutritional aspect 26
    O-linked sugars 37
    protein value 26
    white proteins 37-39, 43, 45
    yolk proteins 37
egg white 19, 105
    alkaline phosphstase
      activity 138
    carbohydrates of 22
    content of IgA, IgM 153
    EGF-like factor 148
    layers of 2
    lipid content of 22
    NGF-like substance 148
    promoting cell growth 146, 148
    ribonuclease 139
    structural ratio of 14
egg white proteins
  (*see* albumen)
    composition and
      physical properties 19
    foaming ability of 128
egg yolk 105
    acetone extraction 81
    acidic phosphatase 137
    alkaline phosphatase 138
    amino acid composition 26
    carbohydrates 18
    cell growth promoting
      activity 146, 147
    cholinesterase 141
    components of 6
    content of IgY 153
    egg yolk powder extraction 80
    emulsifying activity 130, 131
    ethanol extraction 81
    fresh egg yolk extraction 80
    gelation by freezing
      and thawing 130
    glycolytic enzymes 140
    layers of 2
    lecithin 17
    α-mannosidase 139
    minerals 18
    pigments 18
    pyruvate kinase 140
    source of antibodies 152
    structural ratio of 14
    structure of 7
egg yolk immunoglobulin
  (*see* IgY) 103
egg yolk lipid analysis
    flame ionization detection 84
    gas chromatography 84
    HPLC 84
    thin layer chromatography 84
egg yolk lipid fractionation
    ion exchange cellulose
      chromatography 82
    silica gel column
      chromatography 81
egg yolk lipids 74, 81
    acetylcholine 86, 87
    applications 85, 86
    as an antioxidant 90
    as a source of choline 86, 90
    cerebrosides 18
    cosmetics 92
    commercial preparation 80
    dietary cholesterol 85
    dietary fatty acids 85
    liposomes 92
    neutral lipids 75, 78
    phospholipids 17
    sterols 17, 18
    triacylglycerol 74, 75, 80
    tryglycerides 17

egg yolk lipoprotein	
cell growth promoting activity	146
emulsifying properties of	131
egg yolk membrane	105
egg yolk protein	57
amino acid pattern	27, 64
antioxidant action	66
apovitellenin I, II, III, IV, V, VI	59
cholesterol content	67
classes	58
commercial preparation	64
composition and distribution	16
emulsifying properties	62, 63, 131
gelling properties	63
heat stability	64
high density lipoproteins	58
α-lipovitellin	58, 59
β-lipovitellin	58, 59
minerals and vitamins	64, 65
myelin figures (MF)	62
egg white proteins	
(see egg albumen and albumen)	
avidin	19, 22, 31, 39, 148, 181
cystatin (ficin-papain inhibitor)	19, 22, 39, 181
function of	19, 39, 181
lysozyme	19, 21, 39, 119, 181
flavoprotein	19, 21, 39, 181
ovalbumin	19
ovoglobulin	19, 21
ovoglobulin G2	19
ovoglobulin G3	19
ovoglycoprotein	19, 21
ovoinhibitor	19, 22, 39, 181
ovomacroglobulin	19, 21, 39, 181
ovomucin	19, 20
ovomucoid	19, 20, 39, 181
ovotransferrin	19, 20, 39, 119, 181
promoting DNA synthesis	148
eggshell	
calcite crystal	15
disadvantage of washing	182
dolomite crystal	15
eggshell membrane	14, 15
electron micrograph of	3
layers of	2
microflora	181
microstructure of	3
role in bacterial prevention	181
structural ratio of	14
eicosapentaenoic acid (EPA)	32
elastin-like protein	
in shell membrane	5
electrostatic charge	
effect of electrolyte	120
electrostatic repulsive force	
heat denatured OVA	124
embryonic disc	
location in egg	2, 6
emulsified foods	129
emulsifying activity	
whole egg	129
egg albumen	129
egg yolk	129
emulsifying property of egg yolk	
effect of acids or salts	130
effect of drying	130
effect of freezing	130
effect of pasteurization	130
method of evaluation	129
emulsion	
oil in water and water in oil	129
enzymes	
α-amylase	141
catalase	141
cholinesterase	141
found in unfertilized eggs	136
glycolytic enzymes	140
glycosidases	139
nucleoside triphosphatases	139
phosphatases	137, 138
proteolytic enzymes	135

pyruvate kinases	140	folic acid		
ribonucleases (RNases)	139	metabolism	65, 86, 87	
ribonucleic acid-degrading		follicle		
enzymes	140	maturation in ovary	7, 8	
Escherichia coli	43-45, 48, 49, 79, 99	white and yellow follicles	7	
essential amino acids		follicle-stimulating		
suggested daily requirement	27	hormone (FSH)		
estradiol		function on ovulation	7, 8	
on the formation of egg	8	food poisoning		
estrogen		*Salmonella* sp.	182	
effect on hen ovulation	8	symptoms	183	
external thin albumen		*Vibrio* sp.	182	
location in egg	2	free sugar	100	
percentage in egg albumen	19	content in egg	14	

F

G

fatty acid		β-D-galactosidase	136, 139
arachidonic acid (AA)	17, 78, 85, 87, 89	ganglioside	79, 105, 109
docosahexaenoic acid		GD3	105
(DHA)	17, 32, 78, 87, 89	GM3	105
ω-3, ω-6 fatty acid	85, 88, 89	GM4	105
linoleic acid	17, 78, 87	gelation	
linolenic acid	78, 87	effect of pH	119
mother's milk	88, 90	effect of protein	
myristic acid	17, 87	concentration	119
oleic acid	17, 78, 87	effect of salt and sucrose	120
palmitic acid	17, 78, 87	evaluation method of	119
palmitoleic acid	17, 78, 87	heat induced	119
polyunsaturated fatty acid		texture of	119
(PUFA)	17, 88	γ-globulin	
powdered milk	88, 89	in egg yolk	17
stearic acid	17, 78, 87	glucose	14, 18, 22, 100
Fc receptor		β-D-glucosidase	136, 139
binding activity of IgY	161	*N*-glucosidase	140
fertilized egg		β-D-glucuronidase	136, 139
used in virus cultivation	166	glutamate oxaloacetate	
ficin-papain inhibitor		transaminase	136
(*see* cystatin)	19, 22, 39	glutathione peroxidase	166
fish oil	66, 90, 128	glyceraldehyde-3-phosphate	
flavoprotein	19, 21, 181	dehydrogenase	136, 140
foam formation	124	glycosidases	136, 139
foaming ability		glycochemistry	99
effect of pasteurization	127	glycoconjugate	100
influence of freshness		glycolipid	100, 105
of egg albumen	126, 127	α_2-glycoprotein	17, 158
method for evaluation	126	α-D-glucosidase	136, 139
		gonad	7

gram negative bacteria	39, 44, 48	methods for purification	158-160
lipopolysaccharides	39, 48, 49	molecular weight	160
membrane structure	48, 49	number of domains	161, 162
peptidoglycan	48	nutralization titer	
Proteus	49	against rotavirus	157
Pseudomonas aeruginosa	43	prevention of dental caries	173, 174
Salmonella enteritidis	183, 184	prevention of fish disease	171
Salmonella typhimurium	48, 183, 184	prevention of rotaviral	
		diarrhea	168
Shigella	49	productivity of	154
Vibrio cholera	43	specific to α-amidating	
gram positive bacteria	44, 49	enzyme	166
Micrococcus lysodeikticus	49	specific to α_{S1}-casein	162
peptidoglycan structure	46, 49	specific to	
Staphylococcus aureus	44-46	1, 25-dihydroxyvitamin D	166
		specific to *E. coli*	162
H		specific to *E. tarda*	171, 172
hardness		specific to erythropoietin	
heat induced OVA gel	122, 123	receptor	166
heavy metals	80	specific to heat-shock	
hemagglutination	21	protein	166
hematoside (NeuGc)	166	specific to hematoside	
hemocyanin		(NeuGc)	166
hapten	166	specific to human IgG, IgM	166
hexokinase	136, 140	specific to human	
high density lipoprotein (HDL)	16	dimeric IgA	166
emulsifying property	131	specific to human insulin	156, 157, 166
IgM secretion effect	147		
human insulin	166	specific to human	
human neuroblastoma cell	92	transferrin	166
human transferrin	166	specific to mouse IgG	156
humoral immunity		specific to mucin-like	
biological defense system		glycoprotein-A	166
in animals	153	specific to ochratoxin A	166
hydrophobic lipophilic		specific to plasma	
balance (HLB)	85	kallikrein	166
hydrophobic interaction	126	specific to platelet	
L-5-hydroxylysine		glycoprotein	166
in shell membrane	5	specific to prostaglandin	166
		specific to RNA	
I		polymerase II	166
IgY		specific to *S. mutans*	173, 174
(*see* egg yolk immunoglobulin)	180	stability to pepsin	162, 164
content of β-sheet structure	161	stability to trypsin	
denaturation temperature	161	and chymotrypsin	162, 165
differences in properties		structure and function	160-162
to IgG	160-163	oligosaccharides	103, 104, 161
heat and pH stability	162, 163		
illustrative preparation of	152, 154	systemic administration	167, 168

immuno-affinity chromatography	
IgY as a ligand	167
immunogenicity	166
immunoglobulin	
human-human hybridoma	147
in hen's serum	152, 153
of mammals	153
infant formula	67, 88
infection	
in egg infection	183
on egg infection	184
influenza virus	109, 166
infundibulum	
function in oviduct	9, 10
inner shell membrane	
location in egg	3
thickness of	5
inorganic elements	
in egg	15
inositol	65
internal thin albumen	
location in egg	2
percentage in egg albumen	19
iodine value	80
iron-binding proteins	31
isodesmosine	
in shell membrane	5
isoelectric point	
aminopeptidase Ey	137
cholinesterase	141
difference between IgY and IgG	161
glutamylaminopeptidase	137
ovalbumin, ovotransferrin	120
ovomucoid, ovomucin	120
isthums	
function in oviduct	10

K

Kazal-type inhibitor	
ovoinhibitor	22
ovomucoid	20
keratin	
in shell membrane	5

L

lactate dehydrogenase	136
lactation period	110
lactoferrin	42, 44
latebra	
location in egg	2, 6
learning performance	110
lecithin	17, 74
egg yolk	79
rapeseed	79
bovine brain	79
E. coli	79
linseed oil	
hens diet supplement	30
lipids	
composition of egg	28
in eggs	14
nutritional aspect	26, 27
lipoprotein	
growth of embryos	146
liposome	92
lipovitellenin	
(low density lipoprotein)	16, 59
lipovitellin	
(high density lipoprotein)	
α-, and β-lipovitellins	16, 59
liquid eggs	
conditions of pasteurization	187
livetin, in egg yolk	17, 59
allergen in egg yolk	31
α-livetin	17, 59, 61, 103, 158
β-livetin	17, 59, 61, 62
γ-livetin	17, 59, 61, 62, 147
low density lipoprotein (LDL)	16
allergen in egg yolk	31
cell growth promoting	146, 147
emulsifying property	131
LDL1, LDL2	60
lutein	
quantities in egg yolk	18
luteinizing hormones (LH)	
function on ovulation	7, 8
lyochromes	
cryptoxanthin	18
lypochromes	18
lysozyme	19, 21, 38, 39, 45, 136
active site	47, 50

antimicrobial activity	39, 46, 48, 49, 181	neuraminidase	107, 110
		neutral lipids	75
bacteriolytic activity	46, 48, 49	neutral oligosaccharide	100
biological function	48, 49,181	Newcastle disease virus	
conformational transition	47, 49, 50	passive immunization	
cyclic imide	51	using IgY	167
denaturation temperature	119	newcastle disease in hens	152
dimerization	50	nonelastin protein	
dimers	50	in shell membrane	5
disulfide bonds	21, 46, 47	nucleotidase	136, 140
enzymatic activity	45, 46	5'-nucleotidase	136
function of	39, 46, 48	nucleoside triphosphatase	136, 139
hinge-bending	47	nucleus of pander	
modification	48	location in eggshell	2, 6
structure	45, 47, 48,50		
thermal stability	21, 51	**O**	
		oligosaccharide structure	
M		ovalbumin	100, 102
maillard reaction	100	ovomucin	103
mammillary knob layer		ovomucoid	100, 101
location in eggshell	3	ovotransferrin	100, 102
α-mannosidase	136, 139	phosvitin	100, 102
maternal immunity		yolk immunoglobulin (IgY)	103, 104
via placenta in mammals	154	yolk rivoflavin-binding	
maze test	111	protein	103
2-mercaptoethanol		omega eggs	85
resolubilization of		oral tolerance	31
OVA gel	123,124	organic solvents	
microflora		purification method of IgY	158
in the ovary	180	outer shell membrane	
on the eggshell	181	location in eggshell	3
minerals		thickness of	5
content in eggshell	15	ovalbumin	19, 39, 40, 41, 100, 102
content in egg white	15, 22		
content in egg yolk	15, 18	allergen in egg white	31
in each part of the egg	15	carbohydrate chain	20, 38, 40
nutritional aspects	29	denaturation	20, 40, 41
recommended daily		denaturation temperature	119
allowance	30	function	39
molten globular protein	126	mechanism of foaming	129
mouse IgG		mechanism of	
purification using IgY	167	gel formation	124
muramidase	45	ovalbumin A1, A2, and A3	20
		phosphorylation	40, 41, 52
N		supressing cell growth	148
neck of latebra		proteinase inhibitor	40, 41
location in eggshell	2, 6	resolubilization of gel	123,124
Neisseria meningitidis	99	s-ovalbumin conversion	20, 38

serpins homology	40, 41	binding affinity	
structural change		to metal ions	20, 31
on heating	126	carbohydrate moiety	42, 43
structural properties	38, 39	complexes with metal ions	20
structure	40, 41	denaturation temperature	119
structure of oligosaccharides	102	disulfide bonds	44
sulfhydryl-disulfide		function	39, 43
interchange	38	half-molecules	43
sulfhydryl groups		ion-binding	43-45
(SH groups)	20, 38	loctoferrin homology	42, 43
transparent gel of	122, 123	peptides	43, 44
ovary		primary structure	42
bacterial contamination	180	structural properties	43
development of	7, 8	structure	42, 43, 44
oviduct		structure of	
bacterial contamination	180	oligosaccharides	102
protein kinase C	140		
weight and size of	9	**P**	
ovoglobulin		palisade layer	
ovoglobulin G2 and G3	21	location in eggshell	3
ovoglycoprotein	21	thickness of	4
ovoinhibitor	22	passive immunization	
antimicrobial activity	181	application of IgY	166-174
ovokerasin	18	anti-rotavirus IgY	169 170
ovomacroglobulin	21	IgY against *E. tarda*	172
ovomucin	20, 103	IgY against *S. mutans*	174
α-, and β-ovomucin	20	pasteurization	
oligosaccharide	103	condition of	186, 187
promoting cell growth	146, 148	catalase activity	
structure of	20, 21	in egg albumen	141
tumor cell-injuring activity	146, 148	effect on foaming	127
ovomucoid	20, 100	glycosidase activity	
allergen in egg white	31	in egg albumen	139
antimicrobial activity	181	liquid eggs	185
carbohydrate moiety	20	peptidases	135-137
denaturation temperature	119	peptidoglycan	
disulfide bonds of	20	cutting by lysozyme	45, 46, 49
domain I, II, III	20	in membranes	48
structure of		structure	45, 46
oligosaccharides	101	peroxide value (PV)	66, 80
secondary structure of	20	phosphatases	136-138
ovophrenosin	18, 43	acid phosphatase	137
ovotransferrin		alkaline phosphatase	138
(*see* conalbumin)	20, 39, 43-45, 100	phosphatidylcholine (PC)	17, 27, 75-80, 95
allergen in egg white	31	egg yolk	78, 86
antimicrobial activity	39, 43-45, 181	essential nutrient	86
		hydrogenated	92
bacteriostatic activity	20	IgM secretion effect	147

in shell membrane	5	nutritional aspect	
liposome	92	of ω-6 PUFA	32
promoting cell growth	146, 147	pore canal	2
source of choline	86	prostaglandin	89
soybean	78, 86	protein A	
phospholipids (PLs)	74	binding activity of IgY	161
enzymatic conversion	82	protein efficacy ratio (PER)	26
in egg yolk	17, 19	protein-lipid complex	
lysophosphatidylcholine (LPC)	17, 75, 92	cell growth	147
		proteinase	
lysophosphatidyl-ethanolamine (LPE)	17	in egg white	39
		in embryogenesis	52
occurrence of PLs	79	inhibitors	39, 40
phosphatidylethanolamine (PE)	17, 75-77	proteolytic enzyme	135-137
		aminopeptidase	136
phosphatidylglycerol (PG)	76, 77	cathepsin D in egg yolk	136
phosphatidylserine (PS)	76, 77	proteinase in egg yolk	135
phosphatidylinositol (PI)	76, 77	protoporphyrin	
phosphatidic acid (PA)	76, 77, 95	in eggshell	15
polar head group	76	purine N1-C6 hydrolase	136, 140
sphingomyelin (SM)	75, 78	pyrurate kinase	136, 140
structure of PLs	75, 78		
transphosphatidylation	83	**R**	
phospholipase D	82-84	radioallergosorbent test (RAST)	31
cabbage	84		
Streptomyces sp.	82, 84	radioimmunoassay	
phosvitin	16, 31, 60, 100, 102, 138	using IgY	166
		Raman spectrum	
		heat denatured OVA	126
α-, β-phosvitin	61	retinoids	
absorption of iron	61	antitumor activity	28
amino acid composition	61	rheumatoid factor	
carrier of metal ions	17, 61	binding activity of IgY	161
physiological pH	61	riboflavin	17
serine content	16		
structure of oligosaccharides	102	riboflavin-binding protein	
		in egg yolk	17, 103
pigments		ribonucleases (RNases)	136, 139
of egg yolk	18	RNA polymerase II	166
plakalbumin	40	rocket-immunoelectrophoresis	
serpins homology	40	aplication of IgY	166
stereo structure	41	rotavirus	109
plasma kallikrein	166	antibodies specific to	156
polyacryl acid resins	158	diagnosis using IgY	166
polyethyleneglycol	158, 160	pathogen of diarrhea	168
polyunsaturated fatty acids (PUFA)	17, 85		
		S	
nutritional aspect		S-S bond	
of ω-3 PUFA	27	effect of foam formation	129

s-ovalbumin	20
Salmonella strain	182-185
contamination	183
multiplication in shell egg	184
multiplication in liquid egg	185
S. cerro	183
S. enteritidis (SE)	183, 184
S. infantis	183
S. litchfield	183
S. montevideo	183
S. pullorum	183
S. thompson	183
S. typhimurium	183, 184, 186
scaning electronmicrograph	
egg albumen gel	120
shell membrane	5
shell of a hen egg	3
SDS	
resolubilization	
of OVA gel	123, 124
SDS-PAGE	
IgY incubated with	
trypsin and chymotrypsin	165
IgY incubated with pepsin	164
purification of IgY	161
secondary contamination	182
Sendai virus	109
serum albumin	
in egg yolk	17
β-sheet structure	
in denatured OVA	126
shell eggs	184
shell membrane	
amino acid composition of	4
formation of	10
function of	4
lamellar structure of	5
lipid content of	5
location in egg	2
role in bacterial prevention	181
shell stratum	
components of	4
sialic acid	60, 103, 148
sialyloligosaccharide	28, 100
skin feel	92
sodium bicarbonate	10

sodium alginate	
purification method of IgY	158
sodium dextran sulfate	
purification method of IgY	158
sodium sulfate	
salting-out of IgY	159
soy lecithin	91
sphingomyelin	
in shell membrane	4
spongy layer	4
spongy matrix	
in shell stratum	4
Staphylococcus aureus	
protein A	166
stigma	
location in oviduct	9
stradiol	8
Streptococcus mutans	
bacterium of dental caries	173
Streptomyces sp.	84
stress	187
subtilisin	40
suckling mice	
animal model of diarrhea	168
Sudan black	7

T

thick albumen	
effect of foam stability	127
function of	6
location in egg	2, 5
percentage in egg albumen	19
thin albumen	
effect of foam stability	127
function of	6
location in egg	2, 5
percentage in egg albumen	19
tocopherol	91
transovarian transmission	180
transparent gel	
formation of	123
transphosphatidylation	83
tributyrinase	136
tryglycerides	
in egg yolk	17
tumor cells	
injuring factors in hen egg	146, 148
turbidity	
heat induced OVA gel	122, 123

U

ultracentrifugation
   purification method of IgY   158
urea
   resolubilization of
     OVA gel   123, 124
uterus
   location in oviduct   9
   role in oviduct   10

V

vaccine   109
vagina
   location in oviduct   9
   role in oviduct   10
vasotocin (AVT)   10
venom
   nutralization by IgY   168
vertical crystal layer
   location in egg shell   3
   thickness of   4
very low density lipoprotein
(VLDL)   58, 59
   cell growth promoting
     activity   146
vitamin B 12-binding protein   31
vitamins
   bioavailability of   31
   nutritional aspects   28
   recommended daily
     allowance   29
vitelline membrane
   alkaline phosphatase   138
   formation of   10
   glycolytic enzymes   140
   location in egg   2
   nucleoside triphosphatase   139
   structure and components   6

W

white follicle
   maturation in ovary   6, 7
white leghorn
   weight of egg   2, 14
whole egg extract
   promoting cell growth   146

X

xanthan gum
   purification method of IgY   159
xanthophil
   quantities in egg yolk   18

Y

yellow follicles
   maturation of   6, 7
yellow yolk
   accumulation to follicle   8
   deep yelow yolk   6
   light yellow yolk   6
   structure of   6
yolk cathepsins   136
yolk promoting cell growth   146, 147
yolk lipoprotein
   promoting cell growth   146, 147
yolk low density lipoprotein
   promoting cell growth   146, 147
yolk membrane
   receptor specific to IgG   153
yolk riboflavin-binding protein   103
yolk very low density
  lipoprotein
   promoting cell growth   146

Z

zeaxanthin
   quantities in egg yolk   18